Our
Restless
Earth

Our
Restless
Earth

The Geologic Regions

of Tennessee

BY EDWARD T. LUTHER

PUBLISHED IN COOPERATION WITH

The Tennessee Historical Commission

THE UNIVERSITY OF TENNESSEE PRESS

KNOXVILLE

TENNESSEE THREE STAR BOOKS / *Paul H. Bergeron, General Editor*

This series of general-interest books about significant Tennessee topics is sponsored jointly by the Tennessee Historical Commission and the University of Tennessee Press.

Cloth: 1st printing, 1977.
Paper: 1st printing, 1977; 2nd printing, 1985;
       3rd printing, 1991; 4th printing, 1995;
       5th printing, 2003.

This book is printed on acid-free paper.

Library of Congress Cataloging in Publication Data

Luther, Edward T.
Our restless earth.
(Tennessee three star books)
Bibliography: p.
Includes index.

1. Geology—Tennessee. I. Title. II. Series
QE165.L87   577.68   77-21433
ISBN 0-87049-293-4 (cl: alk. paper)
ISBN 0-87049-230-6 (pbk.: alk. paper)

EDWARD T. LUTHER is a native Tennessean whose professional career as a geologist and personal interest in writing have pointed him toward the preparation of this book. Since receiving his advanced degree in geology from Vanderbilt University in 1951, he has come to know the state intimately—first as a team member of the Tennessee Geological Survey and more recently as supervisor of the Survey's research program. He is also an avid reader of fiction and has long been interested in applying writing skills to his technical knowledge in order to make the fascinating science of the earth available to a wider audience.

This book is dedicated to my wife, Pat,

who encouraged me to write about the subject I know best.

Preface

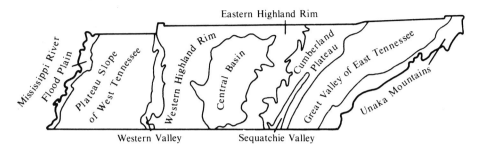

Tennessee is one of the most oddly shaped states in the union, and partly as a result is one of the most diverse. Bordered by eight other states, it is equalled only by Missouri in the number of contiguous neighbors. Though by no means one of the largest states, Tennessee is strung out so far in the east-west direction that Bristol, at the northeast corner, is closer to the Ontario border than to Memphis, at the southwest corner. Memphis, in turn, is closer to the Gulf of Mexico than to Bristol. In fact, the diagonal length of the state from Bristol to Memphis is greater than that of some of the large western states, such as Colorado and Arizona.

Because Tennessee's long axis cuts directly across the grain of the geology, it encompasses parts of several major geologic and physiographic provinces. The land surface ranges from the lofty peaks of the Unaka Range in the east to the pancake flatness of parts of West Tennessee. The rocks are of all sorts, from the once molten granites and altered metamorphic rocks near the North Carolina line to loose sand, silt, and clay near the Mississippi River. The age of the rocks, too, ranges tremendously, from more than a billion years for some to a few hours for those loose sediments just deposited on some sand bar. Most ages in between are also represented. All these factors in combination have produced a rich and varied geologic history, expressed in the subdivision of the state into nine distinct geologic regions, each with a different kind of land surface, underlying rocks, and soil and mineral resources.

This book intends to present the salient facts on each of the geologic regions in a way that will be interesting and comprehensible to the intelligent but geologically untrained reader. Credit goes to Ardi Saint-Clair, former editor of *The Tennessee Conservationist* magazine, for first suggesting a simplified treatment of this subject and to Robert A. McGaw of Vanderbilt University for recommending that the material be published in book form.

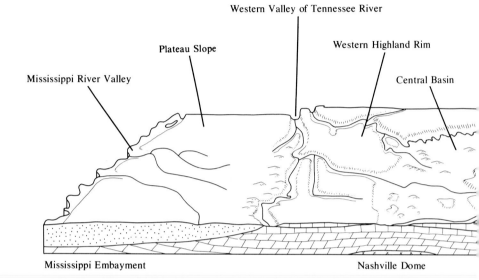

Western Valley of Tennessee River

Plateau Slope

Western Highland Rim

Mississippi River Valley

Central Basin

Mississippi Embayment

Nashville Dome

| Tertiary and Cretaceous sand | Mississippian, Devonian, Silurian, Ordovician, and Cambrian limestone | Ordovician and Cambrian dolomite |

I have chosen to arrange the book in geographical order, beginning at the western end of the state and proceeding eastward; because the complexity of the geology increases toward the east, the reader will thus be introduced gradually to the more difficult material. I have attempted to keep technical terms at a minimum, but there is no way to avoid the introduction of some purely geologic concepts. Only those principles necessary to the understanding of the geology of Tennessee have been treated, however, and this book is certainly not intended to be a textbook on geology.

The professionally trained reader will be aware that I have greatly oversimplified many of the problems, and have stated as fact some theories that still are subject to debate among geologic specialists. If I have erred seriously in any of my interpretations, I can take comfort from the premise that only he who does nothing makes no mistakes. I make no claim to infallibility, but the synthesis of fact and logical inference I have pieced together seems to me to

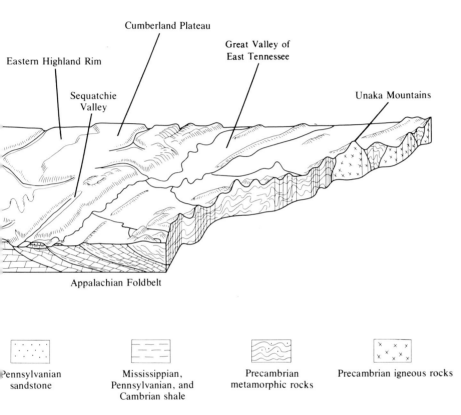

Cumberland Plateau

Great Valley of
East Tennessee

Eastern Highland Rim

Sequatchie
Valley

Unaka Mountains

Appalachian Foldbelt

Pennsylvanian sandstone	Mississippian, Pennsylvanian, and Cambrian shale	Precambrian metamorphic rocks	Precambrian igneous rocks

This relief map (after Miller, 1974) shows Tennessee's major geologic struc-
tures and their relationship to geologic regions of the state.

make a reasonably coherent whole, one that I hope will prove intellectually
satisfying to the reader curious about the origin of familiar landscapes. That
reader must, of course, be the final judge of the success of the presentation that
follows.

EDWARD T. LUTHER

Nashville, Tennessee
May, 1977

Contents

Preface   *page vii*
1. Old Man River's Playground: The Flood Plain of the Mississippi   *3*
2. Home of the Seagoing Dinosaur: The Plateau Slope of West Tennessee   *12*
3. Contrary River: The Western Valley of the Tennessee   *19*
4. Cannonball Country: The Western Highland Rim   *28*
5. Bull's-eye Center of Tennessee: The Central Basin   *36*
6. Plateau of the Barrens: The Eastern Highland Rim   *43*
7. Sky-high Table Land: The Cumberland Plateau   *54*
8. Land of the Rootless Ridges: East Tennessee's Great Valley   *63*
9. Tennessee's Eastern Rampart: The Unaka Range   *75*
10. Explosion Structures: Tennessee's Mystery Craters   *83*
    Selected Readings   *88*
    Index   *91*

ILLUSTRATIONS

Fall Creek Falls   *front cover*
Relief map of geologic formations of Tennessee   *pages viii & ix*
Reelfoot Lake   *4*
Oxbow lakes and meander scars   *7*
Cypress trees in Reelfoot Lake   *7*
Geologic time table   *9*
Mosasaur   *13*
Memphis, viewed from the south   *15*
Exposure of loess in Madison County   *16*
Kentucky Lake   *23*
Examples of unconformities   *25*
Abandoned iron pits near Nunnelly   *30*

Iron-bearing limonite   *33*
Great Western Furnace, Stewart County   *33*
Yearlings grazing on bluegrass   *38*
Short Mountain   *40*
Marine fossils   *42*
Wall of Old Stone Fort   *45*
Duck River Falls   *45*
Morrill Cave   *49*
Oil-drilling rig   *51*
Historic Cumberland Gap   *56*
Rock layers standing on end   *56*
Cumberland Plateau   *59*
Sequatchie Valley   *59*
Natural Bridge, Pickett State Park   *60*
Example of a fault   *64*
Four kinds of faults   *66*
Syncline near Parksville   *68*
Anticline in Douglas Reservoir   *68*
Marble quarry near Knoxville   *71*
Ridges of upper East Tennessee   *77*
Peaks in the Smokies   *77*
Cades Cove   *79*
Drawing of a shatter cone   *85*

Illustration credits: Maps for chapter openings by author. Bulletins of Tenn.
Div. of Geology were sources of the following: pp. viii–ix, 9, 13, 85,
after Robert A. Miller in Bull. 74 (1974); pp. 17, 25, after Ernest E.
Russell in Bull. 75 (by Russell and William S. Parks, 1975); p. 33, after
E.F. Burchard in Bull. 39 (1934); and p. 42, after R.S. Bassler in Bull.
38 (1932). Also Tenn. Dept. of Tourist Dev., pp. 4, 7, 15, 22, 30, 38,
40, 45, 49, 51, 56, 59, 61, 71, 77, 79; U.S. Army Corps of Engr.,
Vicksburg, p. 7; David Royster, Div. of Soils and Geol. Engr., Tenn.
Dept. of Transp., pp. 25, 56, 64, 68; and Phyllis M. Garman, Nashville,
p. 66.

Our
Restless
Earth

1. Old Man River's Playground: The Flood Plain of the Mississippi

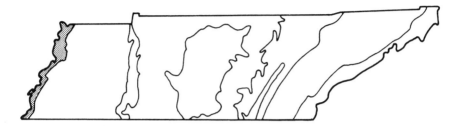

At any given time there must be a worst place in the world to be, and Captain Nicholas Roosevelt found it in the winter of 1811–12, during the maiden voyage of the steamboat *New Orleans*. Steamboats were new, anyway, and the first steam-powered passage from Pittsburgh down the untamed Ohio and Mississippi rivers to New Orleans would have been dangerous even in normal times. If the captain had known that he was sailing straight for the epicenter of the New Madrid earthquake he probably would never have cast off, brave man though he was.

The *New Orleans* made it through, by a miraculous combination of good luck and good seamanship; and her arrival in Natchez, after she was presumed lost, caused great rejoicing and no little amazement. Many others were less fortunate, for the New Madrid earthquake was one of the most powerful and destructive of all recorded history. For a time the *New Orleans* fought her way through tidal waves, past collapsing river banks, and through blockades of floating, uprooted trees. She slid over waterfalls where no waterfalls had ever been, past whirlpools that threatened to drain the whole river into gaps opening in its bed, and even through upheavals that temporarily reversed the flow of the river.

On land the effects were equally devastating. Entire forests were flattened; landslides blocked streams and the few roads that existed in those days; and the earth belched sulfurous vapors, sand, and water from suddenly opened fissures. The ground heaved up in waves that swept across the country like an invading army, occasionally breaking at the crest to form still more fissures. Terrified settlers, to keep from being engulfed, cut down trees to bridge the

gaps and clung to them like life rafts on a stormy sea during the worst disturbances. And finally, great areas of the Mississippi flood plain rose while others sank, forming new lakes as the low places filled with water. Such was the origin of Tennessee's best-known souvenir of the New Madrid earthquake, Reelfoot Lake.

Reelfoot Lake might be termed the jewel of Tennessee's most westerly geologic province, the Mississippi River flood plain. Fourteen miles long and five miles wide, the lake is famous for its eerie, cypress-studded beauty (and also for its fishing and hunting). Beneath deeper parts of the lake the drowned remnants of a heavy pre-earthquake forest still stand. Though perhaps the most famous example, Reelfoot was not the only lake created by the earthquake. Numerous smaller lakes were formed at the same time on the Missouri-Arkansas side of the river. Perhaps even more curious are the areas formerly under water, which are now high and dry. These areas can be identified by stands of cypress trees with fully developed "knees", which are required for breathing by trees growing in open water.

The Mississippi River flood plain occupies an enormous area, extending the length of the river. Although it narrows as it goes northward, it is still about 50 miles wide as far north as Tennessee. The Tennessee portion of the flood plain is narrow, inasmuch as the river, which forms Tennessee's western boundary, flows near the eastern edge of its flood plain in this section. In fact, the area between the river and the Chickasaw Bluffs, which mark the plain's eastern boundary, covers only about 900 square miles of Tennessee. The land is almost completely flat, although, like the river, it slopes southward, dropping about 80 feet from the state's northern boundary to Memphis. Breaking the flatness to some extent are low ridges, built by the river, that form natural levees along its present banks as well as along its innumerable abandoned courses. The comparatively slight swells and depressions made by the New Madrid and earlier earthquakes also offer some variety. Between the natural levees are scroll-like, oxbow lakes and meander scars. These features show that every square inch of the vast flood plain has been occupied, probably several times, by the river, which has looped and swung its way from side to side, its serpentine course shifting gradually downstream like a sidewinder rattler. Each loop of the river forms an almost complete circle, returning to a point separated only by a narrow neck of land from its starting

Reelfoot Lake is the state's best-known souvenir of the New Madrid earthquake. Drowned remnants of a pre-earthquake forest still stand beneath the surface of deeper parts of the lake, but the visitor today sees mainly the beauty of cypress trees, with their weirdly shaped "knees" elevated for breathing.

point before twisting away in the opposite direction to form another loop. In time of flood the river cut across the necks of the loops, temporarily shortening its course. Later, new loops formed, constantly shifting the course and moving it from side to side.

Over countless centuries the Mississippi and its tributaries have been carrying toward the Gulf of Mexico a vast amount of material eroded from the central part of the North American continent. Much of this sediment, an alluvial blanket, has been deposited on the rivers' flood plains. Sand, gravel, and clay from the east slope of the Rocky Mountains and the west slope of the North Carolina Blue Ridge are mingled within it. At some places the blanket is several hundred feet thick, and all of it is saturated with water, at least to within 20 feet of the ground surface. The great thickness of water-soaked alluvium accounts, at least in part, for the devastating effects of the earthquake. Such material has little strength, and when subjected to the alternating compression and relaxation of earthquake-generated shock waves, it behaves like a gigantic bowl of jello. Vibrations that would be felt in hard-rock areas as no more than momentary tremors were enough to turn the water-soaked alluvium to quicksand and set up wavelike motions of the ground.

In explaining why the effects of the earthquake were so great on the flood plain, new questions arise almost as fast as the old ones can be answered. Why is alluvium so thick in this flood plain that its base is well below sea level in places? And why were the earthquake epicenters aligned parallel to the course of the river? And, finally, what were the mechanisms that caused the earthquakes themselves?

To answer these questions we must go back millions of years and, in imagination, peel back the successive geologic events like the layers of an onion. We must go back far enough in geologic time to examine the hard rocks below as well as the unconsolidated sediments at the surface of the Mississippi flood plain.

So that the reader can understand some of the terminology the geologist uses to speak of the various subdivisions of geologic time, it might be useful to explain some of the more obscure terms. Geologic time is vast, measured in billions of years, and for many events there is not enough information to assign accurate dates in years. Such events are commonly assigned relative dates; i.e., they are said to have occurred before some events, after others. In

Abandoned courses of the Mississippi show up (above) as oxbow lakes and meander scars in a high-level photo of the modern-day river, and (below) as rows of cypress in Reelfoot Lake, which grow along low ridges that bordered meanders existing before the lake was formed.

order to introduce more precision into these relative dates, geologic time has been subdivided into major units called eras (see Fig. 1), each of which, in turn, is subdivided into periods. The eras lasted for hundreds of millions of years, the periods for tens of millions.

In general, much less is known about the older eras and periods than the younger, because the events of more recent time tend to obscure the evidence that would permit us to interpret the older periods. The oldest rocks exposed at the surface over most of Tennessee were formed during the Paleozoic era (a term derived from Greek root words meaning "time of ancient life"). This era began about 600 million years ago, lasted for 370 million years, and is subdivided into seven periods, the oldest of which is called the Cambrian. Because so little is known of earlier time, rocks originating before the beginning of Paleozoic time are commonly lumped together under the catch-all term, Precambrian. This rather cavalier treatment condenses nearly four billion years of earth history into one unit. To be sure, the Precambrian era (or eras) is subdivided in various ways at various places, but there is as yet no general agreement about the best way to do it.

Following the Paleozoic era came the Mesozoic era ("time of middle life"), which lasted 162 million years and includes three periods, of which only the last (the Cretaceous) is represented in Tennessee. By now it becomes apparent that the system for naming eras derives from a description of the kind of life that existed during each respective time, so the term Cenozoic ("time of recent life") for the most recent era follows naturally in the sequence. The Cenozoic era, which consists of only two periods, has lasted from about 68 million years ago until the present.

Between five and six hundred million years ago (during the last part of the Cambrian period), the area was the bed of a shallow sea, and far from land. For perhaps 225 million years it remained so, and great thicknesses of limy ooze were precipitated from the sea water in layers that hardened into solid limestone and dolomite. These layers of rock, and the sea that deposited them, covered much of the interior part of what is now the United States. Today these rock layers are deeply buried in West Tennessee, but they appear at the surface in Middle and East Tennessee.

At some time during the vast span of years since the earliest marine layers were laid down, forces from deep within the earth pushed the limestone and dolomite layers up into broad, low domes (the Nashville Dome to the east and the Ozark Dome to the west), connected by a low arch called the Pascola Arch. Periodically renewed uplift has kept the domes high, but not the connecting arch. About 325 million years ago the entire area rose above sea level, and millions of years of erosion removed the crest of the arch. During

ERA	PERIOD	YEARS BEFORE PRESENT
CENOZOIC	Quaternary	2,500,000
	Tertiary	
		68,000,000
MESOZOIC	Cretaceous	
		140,000,000
	Jurassis	
		205,000,000
	Triassic	230,000,000
PALEOZOIC	Permian	
		285,000,000
	Pennsylvanian	
		325,000,000
	Mississippian	350,000,000
	Devonian	
		410,000,000
	Silurian	430,000,000
	Ordovician	
		500,000,000
	Cambrian	
		600,000,000
PRECAMBRIAN		

This generalized table of geologic time indicates the consecutive eras and periods and their relationship to the present.

the latter part of the Mesozoic era a new pattern of crustal warping began to dominate, and the crust sagged along a line at right angles to the east-west line of the older Nashville Dome, Pascola Arch, and Ozark Dome. About 75 million years ago, during the Cretaceous period, the sea, again invading up the lowest part of the downwarped area, had formed an arm of the Gulf of Mexico called the Mississippi Embayment that reached as far north as Cairo, Illinois, and had begun depositing thousands of feet of loose sediment on top of the limestone and dolomite already in place. Even today these younger deposits are largely unconsolidated; that is, they have never been compressed and cemented into hard rock as have the older sediments. Eventually over 3,000 feet of sand, silt, and clay buried the remains of the Pascola Arch, which still crossed the embayment from one side to the other like the bent and cracked but still tough ridgepole of a sagging barn roof. To visualize the difficulty of warping such a resistant mass, try bending a flat magazine and then see how much harder it is to bend one that is rolled up. Cross-folding the Pascola Arch, in fact, proved so difficult that it ruptured instead, along a line parallel to the trough of the embayment, and, incidentally, parallel to the river. Warping down of the embayment continues even today, causing periodic movements along this buried fault. An unusually large or sudden movement, or series of movements, on the fault in 1811–12 doubtless generated the shock waves of the New Madrid earthquake.

The origin of the Mississippi River itself dates back more than 40 million years. When the bottom of the Mississippi Embayment began to rise, and the sea retreated to the south, the ancestral Mississippi River formed, draining the lowest trough of the emerging embayment, and its length has gradually extended headward and seaward ever since. From the beginning the river drained a very flat area, so it must have begun to build its flood plain almost immediately.

The great thickness of the alluvium must, however, be ascribed indirectly to the effect of the great glaciers that, during the last ice age, covered much of North America to a depth of thousands of feet. True, these continental glaciers never advanced as far south as Tennessee, stopping at about the present course of the Ohio River (this is no coincidence, since the boundary of the ice sheet determined the river course). But the glaciers indirectly affected the south in a number of ways. By locking up much of the earth's water in the form of ice, the sea level was lowered several hundred feet. At the same time, the enormous weight of the ice depressed the northern half of the continent, and in compensation the southern half stood higher. As the downward push in the north continued, the Mississippi drainage area stood higher and higher above the Gulf of Mexico. The river automatically responded by cutting its valley downward in a deep gorge to match the new level of the sea. The gorge, several hundred feet deep, extended far up the river, past Tennessee, and up the Ohio and other tributaries. Then, when the glaciers melted, the process reversed: the sea level rose, drowning the mouth of the Mississippi, and the now unburdened northern part of the continent began to rise to its former position. Erosion went to work on the rock and soil that the glacier had bulldozed to the south, and used it to fill up the gorge so laboriously dug by the river. No surface evidence of the gorge remains, and only from holes drilled to test foundations for dams, locks, and bridges do we know of its existence. As a final contribution, the retreating glacier uncovered, along with the rock and soil, tremendous quantities of finely ground rock flour. As this material dried, it was picked up by the wind in dust storms that, by comparison, must have made those of the 1930s dust bowl seem like a clear day. The Chickasaw Bluffs that form the eastern edge of the flood plain are capped with deposits of this wind-blown rock powder (loess) measuring as much as 80 feet thick, some derived from the glacial outwash and some from the silt component of the flood plain's alluvial blanket.

And so we see that for 550 million years, since the middle of the Cambrian period, nature has been building a mechanism that can generate earthquakes. The mechanism consists of a comparatively rigid layer below, gradually being bent and broken by the weight of thousands of feet of sediment above, which in turn constitutes the ideal medium for transmitting the shock waves generated by the breaks. But what are the chances of our seeing a repeat performance of Mother Earth's belly dance of 160 odd years ago? With luck, perhaps none. Scientific studies suggest that such powerful quakes occur, on the average, no oftener than once in a thousand years, and we've had ours for this millennium, thank you. Or, on the other hand, perhaps not.

2. Home of the Seagoing Dinosaur: The Plateau Slope of West Tennessee

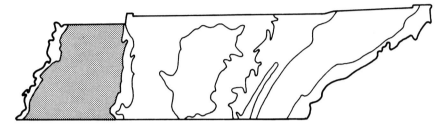

The pearl-gray surface of the water rippled away to infinity, shading into misty, early-dawn pinkness at the horizon. As the light grew stronger the mist lifted into tendrils and patches, suggestive of eerie moors and uncanny creatures. The water bulged upward, and a head appeared above the surface, supported by a long, serpentine neck. Higher and higher rose the head, until its huge slash of a mouth, filled to overflowing with viciously pointed teeth, reached a full fifteen feet above the water's surface. Ponderously the creature surged forward, its barrel-shaped body propelled by flipper-like limbs and a long, lashing tail.

Nearby, the surface of the water boiled as a school of fish broke the surface, only to be scattered by another form, more streamlined and shorter of neck than the first but equally blessed with teeth, that hurtled through the school and seized an unlucky victim. Before the school could escape, the long neck of the larger monster flashed like a cracking whip, and he too seized his meal from the sea.

Our imaginary scene is laid not at Loch Ness but at Coon Creek, in McNairy County, one of the most famous fossil-collecting localities in the United States. Seventy million years ago, near the end of the Cretaceous period, this part of West Tennessee was covered by a shallow sea that abounded in marine life—fish, shells, great marine turtles, and the giant seagoing dinosaurs (actually giant reptiles, but closely related to the dinosaurs). By this time the plesiosaurs, those long-necked, sixty-foot leviathans, were well on their way to extinction, and in fact, placing one at the Coon Creek scene requires a little poetic license, since no remains of them have been found there. They did exist, though, for their skeletons have been unearthed at many places in the west in rocks of the same age, and their distant cousins, the mosasaurs (the short-necked predator of the opening scene) have left their forty-foot skeletons, or at least scattered parts of them, in the bone beds at Coon Creek.

Materials for spectacular restorations of the great vertebrate reptiles that lived and died here are not the only important contributions of the Coon Creek fossil beds. Of equal, or perhaps even greater, scientific significance are the myriads of shells of clams, oysters, crabs, and other ancient sea dwellers, so beautifully preserved that their close kinship to living salt-water animals is immediately evident. It is all very clear cut and logical, until you ask yourself how these sea dwellers came to leave their remains more than 400 feet above sea level in West Tennessee.

The sea that nourished the strange creatures found at Coon Creek was also responsible for features that make the Plateau Slope of West Tennessee, one of Tennessee's major geologic provinces, different from the rest of the state.

The mosasaur depicted in this rough sketch was a sea-going monster that sometimes reached 40 feet in length. Its remains have been found in the fossil beds at Coon Creek in McNairy County.

Extending from the Chickasaw Bluffs overlooking the Mississippi River flood plain to the western valley of the Tennessee River and north from the Mississippi state line to Kentucky, the province includes nearly 9,000 square miles.

Most people think of West Tennessee as being as flat as the palm of your hand, and to describe it as a plateau might raise a few eyebrows, but it is nevertheless true. At its eastern margin it stands more than 300 feet above the Tennessee River, and though it slopes gradually to the west, the western edge is still nearly 200 feet above the Mississippi.

Actually, it might be easier to get a mental picture of the area by considering it to be two provinces instead of one. To the east, forming a belt from ten to thirty-five miles wide parallel to the Tennessee River, are the sand hills. To the west of the sand hills region the rest of the province is the truly level part of West Tennessee, a broad, flat-topped area between the headwaters of the streams that drain eastward into the Tennessee and those that drain westward into the Mississippi.

The difference in topography between the eastern and western parts of the Plateau Slope is caused, for the most part, by differences in the kinds of rocks to be found in the two regions. This might sound a little odd, because most people would say there are no rocks in the region at all. To a geologist, however, a rock does not have to be hard. Any layer of material—sand, clay, volcanic ash, salt, or similar deposit—that is thick enough and extensive enough to make up a sizable unit of the earth's crust is by definition a rock, even though it may not be compacted or cemented together, as for instance in sandstone. In this context the thick layers of sand, silt, and clay laid down by the sea 70 million years ago are rocks, as are the similar layers deposited in younger seas during the next 30 million years, and even the thick layers of wind-blown dust (loess) that were deposited at the close of the last ice age, about 10,000 years ago. It is these layers of loess that cap the western part of the Plateau Slope and account, in large part, for its flatness.

We have all heard the term "terra firma", with its implication of the everlasting solidarity of dry land, but the existence of places such as the Coon Creek fossil locality belies the whole concept. After all, where Memphis now stands was once open sea, and where the mosasaur swam in unfettered freedom is now a cotton field, nearly 350 miles from the sea. But there is more: earlier still, before the sea overwhelmed the land 70 million years ago,

As seen in this view from the south, Memphis is built on a flat upland surface of loess, deposited about 10,000 years ago, that caps the underlying bedrock of the western part of the Plateau Slope.

mountains stood where now the Mississippi flows. And before that? Perhaps it would be less confusing to approach the geologic history of the region in chronological order, or at least as much of it as can be deciphered.

From about 550 million years ago, during the Cambrian period, until about 325 million years ago, at the end of the Mississippian period, the seas rolled, life evolved, and layer after layer of rocks were deposited in West Tennessee as described in chapter 1. The seas were shallow, and by about that time areas to the north and east became filled with sediment and were built up above sea level. It was during this same period that the Ozark Dome, the Pascola Arch, and the Nashville Dome, mentioned earlier because of their influence on the history of the Mississippi River flood-plain region, came into being. During the interval from 305 million years ago until more than 70 million years ago, the continued upward push of these structures kept the entire West Tennessee area above sea level. During at least part of this time the area now occupied by the Mississippi River flood plain and adjacent parts of the Plateau Slope probably stood more than 4,000 feet above sea level, higher than any point in Tennessee today west of the Smoky Mountains.

During this entire time span of about 235 million years, the Plateau Slope was relentlessly rasped by the erosive force of strong winds, storms, and flowing streams, until the domes and arches were beveled off almost to their nethermost roots. The Pascola Arch, which had been a semidome, was now planed off like a pine knot in a two-by-four, and the whole country was reduced to a nearly featureless plain.

The next chapter in the geologic history of the Plateau Slope could be entitled "The Return of the Sea." Where earlier, uplifting forces had created domes and arches, now in the ensuing epoch the earth's crust sagged into a trough through which the sea made its way as far north as Cairo, Illinois. Somehow the support deep beneath the earth's crust gave way, and all of western Tennessee, Kentucky, Mississippi, and eastern Louisiana, and all of eastern Missouri, Arkansas, and western Louisiana bent downward into a north-south trough that cut across the preexisting east-west structures of the domes and arches. And as soon as the sea finished its reoccupation of this part of its ancient domain, washing bare the underlying bedrock as it advanced, the deposition of marine sediment began again, but with a difference. This time some nearby areas remained above sea level, and sediments that had eroded from the continental land mass began immediately to fill the new embayment,

A road cut in Madison County exposes a six-foot layer of loess, the wind-blown dust that was deposited on top of sand left by the Tennessee River or its tributaries long ago, when they flowed at a much higher level than now.

like scar tissue closing a wound. At the same time that the bottom was subsiding, clay, silt, and sand washed into the shallow sea and filled it. By about 40 million years ago, when the Tennessee portion of the embayment was finally filled by several thousand feet of sediment, the sea again retreated southward toward the Gulf of Mexico.

All time and nature seem to be in a continuous state of flux, and the expulsion of the sea from West Tennessee was by no means the end of change. A new and equally remarkable, though much briefer, time had begun—the Pleistocene ice age. Still unexplained climatic changes led to a vast expansion of the polar ice cap, together with the formation of other centers of ice accumulation, and most of the northern half of North America (as well as Europe and Asia) was covered with a coalescing series of ice sheets. The ice thickened enormously, to several thousand feet, in fact. As happened in the development of the Mississippi flood plain, the lowering of sea level resulting from the transfer of water from sea to glacier caused every stream that still flowed free of the ice sheet to cut its valley deeper. The Tennessee and all the smaller streams that drained the Plateau Slope developed gorges far below their present levels that were filled again with sediment when the ice retreated, about 10,000 years ago. As a final smoothing over, like icing on a cake, the area south and east of the glacier's maximum advance was buried by the layer of loess that caps the Mississippi River bluffs from the Ohio River south to Vicksburg, Mississippi, and beyond. Since that time the countryside seems to have been subjected only to those forces and events that can still be seen acting on the land.

The 550 million years that have been examined in the Plateau Slope of West Tennessee represent only a small fraction of the time that the earth has been spinning through the universe, estimated at 4.5 billion years, but evidence of what happened during the most distant part is missing. The earliest records available to us are in the sedimentary rocks, which show that during the first half of the last 550 million years the region was under salt water. And during the last half of this period, the sea played a return engagement for 30 million years or more. It was only after this last marine interlude was finished that uplift and erosion shaped the present Plateau Slope of West Tennessee.

3. Contrary River: The Western Valley of the Tennessee

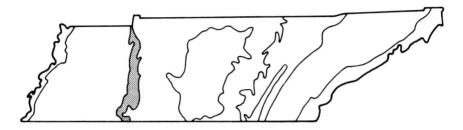

The American Revolution was dragging toward the end of its fifth year, and the battle of Kings Mountain was nearly a year in the future when John Donelson and his flatboat flotilla set out from Fort Patrick Henry on the Holston River enroute to a rendezvous with the Robertson party in the unknown territory of Middle Tennessee. December 1779 was a treacherous time to head for the unknown, and the perils and trials of that journey tell us much about the kind of people who opened the Tennessee frontier, and, incidentally, about the territory into which they moved.

The aptly named *Adventure*, flagship of the fleet, was frozen in the ice, stuck on mud banks, whirled about in a treacherous current called "The Suck," attacked repeatedly by Indians, and tossed by rapids like a chip in a mill race, all on the southwestern and northwestern stretches of the Tennessee River. By contrast, after the river turned north the few shots fired by Indians from the distant banks did little but break the monotony of the smooth, uneventful passage north to the Ohio. Why was that part of the river so different from the other?

The free-flowing Tennessee of the eighteenth century had ample reason for its numerous peculiarities. After all, it could be said, without stretching the truth too much, that the river flows across or beside every one of the varied geologic regions of the state except the Central Basin and the Mississippi River flood plain. Near its headwaters the tributaries of the youthful Tennessee come swooping steeply down the slopes of the Unakas. It grows into a major river while zigzagging southwestward through the ridges and valleys of East Tennessee. Showing mature strength, it cuts through the mountainous barrier of Walden Ridge and the Cumberland Plateau in a gorge so steep and wild it is sometimes called the Grand Canyon of the Tennessee. This was a scene of travail for the Donelson party when a gigantic swirl, caused by the constriction of the stream bed in the canyon and the eddy of an inflowing mountain tributary, threw the flotilla close to shore, where the settlers were vulnerable to Indian attack.

Farther downstream they came to the perilous white-water stretch known as Muscle Shoals, representing a nightmarish three-hour fight against death by drowning for the Donelson party. The dangerous shoals were created by the efforts of the river to cut through a tough, resistant layer of rock now called the Fort Payne Chert.

But from Muscle Shoals, near the common corner of Tennessee, Mississippi, and Alabama, north to the Ohio River there was little to impede progress. The shoals are the last barrier to navigation caused by rock outcrops in the stream bed. From there to the Ohio, the Tennessee flows north on alluvium; and since the material in the river bed is soft and unconsolidated, the gradient is smooth and uninterrupted, hence no rapids or waterfalls. The chief obstacles to navigation were shifting channels, wandering sand bars, and the ever treacherous snags, floating trees, and "sawyers" (floating trees anchored at one end in the river bed, so that the free end rose and sank with fluctuations in the current).

Not least among the peculiarities of the Tennessee River is the fact that it ends up flowing north. Why does it behave so irrationally? A normal river, once it has gathered enough water from inflowing tributaries to establish itself as a full-time businesslike stream, heads by the most direct course available down the slope to the nearest ocean. But the Tennessee defies such logic.

To be sure, when the river starts out as a system of small streams draining the west slope of the Blue Ridge in North Carolina and Virginia, it seems purposeful enough, moving off briskly toward the southwest and the distant Gulf of Mexico. But contrariness sets in about halfway there, near Chattanooga, where the river jinks to the right like a broken-field runner and breaks through thousand-foot-high Walden Ridge to Sequatchie Valley. Thereafter it resumes its Gulf-ward course—but not for long. Near Guntersville, Alabama, the river again kicks over the traces and heads off to the northwest, away from, not toward, the Gulf, which is the goal of other mid-continental rivers.

Finally, however, after fighting its way through Muscle Shoals, the river makes a last, decisive turn away from the Gulf and heads due north across Tennessee, to empty its waters into the Ohio at Paducah, Kentucky. This northward reach is the beautiful Western Valley of the Tennessee, another of the state's major geologic regions.

The northward-flowing Tennessee is a different sort of river from the brawling, turbulent one of the upper reaches, or at least it was before the entire river was tamed by a series of TVA dams. To a modern voyager most of the river hazards are no longer apparent, the differences btween the upper and lower river having been obscured. The Tennessee is now a series of man-made lakes. Dams keep pool levels high enough to provide a nine-foot channel right over Muscle Shoals and all the other obstructions so dreaded by early river men, and locks lift boats from one pool level to the next. The Western Valley of the Tennessee is now perhaps better known as Kentucky Lake.

Perhaps the best explanation for the strange course of the Tennessee River is that probably it is not a single river at all but rather is made up of at least three earlier rivers, stuck together like the parts of some Rube Goldberg contraption designed to move water from one place to another in the most unlikely way possible. By far the oldest of the three originally separate streams is the segment that now constitutes the upper (East Tennessee) reaches of the river. Many geologists believe that millions of years ago this part connected with the Coosa River in Alabama and drained into the Gulf of Mexico at Mobile Bay. The development of that part of the Tennessee drainage must have begun as soon as East Tennessee rose above sea level, about 285 million years ago.

The northwestern reach of the river was the second to develop. It probably originated 75 million years ago as a sluggish stream flowing northwestward across the coastal plain and into the Mississippi Embayment at a point somewhere in northeastern Mississippi. From a tidal creek it grew steadily headward as more and more of the drainage of Mississippi and Alabama was diverted into this short cut to the sea, until it finally reached far enough to the southeast to capture the drainage of the upper Tennessee River and divert it toward the northwest. Then, about 40 million years ago, the sea retreated southwestward from the Mississippi Embayment toward its present boundaries, and the ancestral Tennessee River must have curved away to the southwest in order to follow along behind, perhaps connecting for a time by way of Bear Creek with the present Tombigbee River drainage.

About this time also the third and last segment of the Tennessee River system, the northward reach, had its beginning. The excavation of the north-south valley began, but the stream that cut it at first flowed south, to the Tombigbee connection. One of the most interesting lines of evidence for this

relates to the angles at which many of the tributaries, particularly the Duck River, approach the Tennessee. River confluences ordinarily form acute angles pointing downstream, and it would thus seem reasonable for tributaries flowing into the northward flowing Tennessee to flow northwestward or northeastward. Instead, many of them flow toward the Tennessee at angles pointing south, as though the master stream had flowed that way at the time the courses of the tributaries became established. Only when nearing the Tennessee do most of the tributaries turn abruptly and angle in toward the north. This may well represent an adjustment that the stream courses have made since the direction of flow of the Tennessee reversed, from south to north.

If this reconstruction of the history of the Tennessee River is correct, the next major event was the reversal of flow direction of the Tennessee's western segment, brought about indirectly by the Pleistocene ice age. As described in discussions of the Mississippi River flood plain and the Plateau Slope, the lowering of sea level during the ice age caused the valleys of the Mississippi and Ohio rivers to be deepened enormously. The gorge of the Ohio River was cut below the level of the valley of the Tennessee, so that water from the valley began to flow northward into the Ohio, beginning the erosion of a new gorge. This gorge ate its way southward along the valley of the Tennessee, diverting the drainage bit by bit. When the head of the gorge reached the Alabama section of the old river it captured the drainage, thus establishing the present course of the river. After the ice sheet melted, sea level rose again, and the Tennessee, like the Ohio and Mississippi, filled its Pleistocene gorge with sediment, but has retained the anomalous course imposed on it by this long and complicated history.

The geology of the Western Valley is as complicated as the history of the stream that cut it. In order to understand it, one must realize that the rocks exposed in the valley and on the highlands to either side record the events of many millions of years, and belong to at least three distinct and different eras of the earth's history. In some places rocks belonging to each of the three eras can be seen in sequence, one above the other; in some places rocks representative of only one, or two, of the eras are present. Because of this, deciphering the geologic history of the region is like trying to put together a jigsaw puzzle, only to discover that the pieces belong to three different puzzles.

To show how the rocks of one period of the earth's history are related to

Kentucky Lake now occupies much of the Western Valley of the Tennessee River. This view looks across the lake from Paris Landing State Park, on the west side of the valley.

those of another, it is necessary to introduce here a strictly geological term, the *unconformity*. An unconformity is an irregular surface, resulting from erosion, that separates rocks of one age from those of another. It is formed when the conditions that lead to the deposition of one kind of rock come to an end, and a period of time elapses before more rocks are deposited on top of the earlier ones. Most commonly, during such an interval, erosion takes place, removing some of the earlier deposit and leaving an irregular surface on which later rocks are laid down. In other cases, not only does erosion take place, but also, particularly if the time interval is very long, some disturbance in the earth's crust may fold or tilt the layers of rock, so that when sedimentation resumes, the rocks laid down above the unconformity may lie flat across the edges of sloping beds below the unconformity.

It is a measure of the complexity of the geology of the Western Valley of the Tennessee that no less than ten unconformities have been recognized, ranging from the very minor and obscure to a number of well-marked and prominent ones. When you consider that the erosion accompanying the formation of any unconformity can cut down through earlier unconformities and their pre-existing rocks, then a younger bed could rest at different places on various older beds. This is indeed the case in the Western Valley.

The oldest rocks exposed in the Western Valley were deposited from 450 to 325 million years ago, during the early Paleozoic era. For most of this time shallow seas, discussed earlier, covered the entire eastern part of the country, advancing and retreating as the crust of the earth rose or sank. During the times when the sea was most extensive, and shore lines were far away, layer after layer of limestone and dolomite were deposited on the sea floor, much like the deposits being laid down on the floor of the Caribbean Sea today. At other times the sea became shallower; shore lines advanced toward the south and west; and mud, silt, and sand from the land areas were swept into the sea to mingle with the carbonate sediments of the open ocean. A number of times, apparently, the area was lifted entirely above sea level, and erosion began to remove the marine sediment. Each such emergence of the area gave rise to an unconformity, representing a gap in the geologic record during which rocks were removed rather than deposited. But after each emergence the sea returned, and new marine layers buried the old. The youngest of these now completely hardened rocks preserved in the Western Valley were deposited

The horizontal, darkish line in each of these exposures represents an uncon-formity—the irregular surface of an older bed of rock that was eroded before later seas deposited sediments for other rocks. Above: I-40 in Putnam County. Below: Pine Top railroad cut, Hardeman County.

about 325 million years ago, and so we know the cycle of marine invasions and retreats continued at least until that time.

Probably it went on much longer, but at the top of the hard-rock layers in this area is a very profound unconformity, and the beds that may once have been present above have been stripped away by erosion. From the fact that the older rocks have been compacted and cemented into hard rock, it is evident that they must have once been buried to a considerable depth. This reasoning is supported by the "youth" of the softer rocks that rest above the unconformity, only about 75 million years old, as determined by their fossil content. The unconformity thus represents an interval—a gap in our knowledge—of as much as 250 million years, during which erosion could have removed thousands of feet of beds. We cannot know, of course, exactly when the sea withdrew and erosion commenced, but it seems reasonable to assume that a large part of the missing time interval was once represented by rock layers that have since been eroded away.

The geologic record in the Western Valley takes up again with the deposits laid down from about 75 to 40 million years ago. These are layers of sediment that are yet to be cemented into hard rock. They represent a new marine invasion that followed the long period of erosion occurring while the area was above sea level. This marine invasion had begun along the Gulf Coast almost 200 million years ago, and then had inched its way northward until, late in the period, it swept across Tennessee. Toward the end, the northward invasion was accelerated by the sagging of the earth's crust into the trough of the Mississippi Embayment. When this arm of the sea reached its maximum size it covered all the area of the Mississippi River flood plain, the Plateau Slope, and the Western Valley, and extended some distance on to the east into the Western Highland Rim region.

The deposition of the soft sediments that still blanket West Tennessee was completed about 40 million years ago during the Tertiary period, when the sea withdrew from the embayment for the last time. There seems to be no record of what happened in the Western Valley—or anywhere else in the state, for that matter—for almost 30 million years thereafter. It must have been almost totally a time of erosion, and the material eroded from Tennessee probably exists today as part of the Gulf Coastal Plain or continental shelf.

At any rate, the next geological evidence available indicates that at least part of the Western Valley was indeed a valley as long ago as 5 million years or more. Deposits of river alluvium, particularly in the southern part of the valley, high above the level to which the river has since cut, provide the clue. These deposits, which may be related to the earlier southwest-flowing part of the Tennessee, rest on terraces cut by the river when its bed was several hundred feet higher than at present. They drape off the upland surfaces, down the gentler slopes of the valley sides, and merge imperceptibly with lower terrace deposits laid down during the ice age, and in some cases even with the alluvial deposits still being created by the river today. Mapping these deposits shows where the river was before the ice age, and the fossils found in these and the ice age gravels reveal a fascinating picture of a countryside populated by camels, prehistoric elephants, giant sloths, dog-sized horses, and saber-toothed tigers, to mention only a few of the more bizarre inhabitants.

Mother Nature, it seems, does not believe in halfway measures. Not content with draining away the sea to the Gulf, she followed up with the ice age that took so much water out of circulation that sea level dropped several hundred feet below present levels. Then ensued the gorge-cutting that created the northward flowing Tennessee; and, after the close of the ice age and the rise of sea level, came the filling of the gorge and the deposition of the river-bottom alluvium of the present erosional cycle. And so it seems that all nature has conspired, for millions and millions of years, to twist the tail of the Tennessee River until that tail almost approaches the head. Still, in the long view of geologic time, the present course of the Tennessee is surely a temporary anomaly. Given a few million years of stable geologic conditions, the streams flowing south across the coastal plain into Mobile Bay will continue to cut headward; and very probably will again pirate away the Tennessee and return it to its original course.

However, man, the modern meddler with ancient geologic processes, is not likely to sit back for millions of years and let nature take its course. One of the odder quirks of fate is that the most up-to-date scheme for improving the Tennessee—the Tennessee-Tombigbee Waterway—would in a sense restore a former course of the Tennessee, abandoned now for a million years. As the old saying has it, the more things change, the more they remain the same.

4. Cannonball Country: The Western Highland Rim

On January 8, 1815, when Sir Edward Pakenham's British army attacked New Orleans they did not know that General Andrew Jackson's army was entrenched across their path. When Jackson's Tennessee marksmen laid down a devastating barrage of rifle and artillery fire, Pakenham followed his European training, closed ranks, and fed his troops into the hail of death like a log into a sawmill.

The "Unnecessary Battle," so called because it took place in ignorance that a peace treaty had already been signed, ended the fighting in the War of 1812. Jackson's sharpshooters (with a generous assist from Jean Lafitte's privateer artillerymen) salvaged a measure of national pride from that disastrous war, and eventually put Jackson in the White House.

Another Tennessee contribution to that victory is less well known. Iron mining, smelting, and fabrication in Tennessee date back to the earliest years, and by 1810, Montgomery Bell had established in Dickson County a furnace for smelting locally mined ore. From this furnace came the cannonballs used by Jackson's artillery.

The iron industry of that part of Tennessee has been defunct for many years. Few people even seem aware there was such an industry, but in fact it flourished in both Middle and East Tennessee. Remains of the old furnaces still stand at many places, stark monuments to an almost forgotten prosperity in areas now largely gone back to wilderness. Other reminders abound, in place names like Iron Hill, Furnace Hollow, Upper Forge Branch, and many others—all related in some way to a vanished industry of the mid-1800s. For the most part, in Middle Tennessee the industry was restricted to what is now known as the Western Highland Rim, another of our major geologic regions.

Why the industry rose here, flourished, and died, provides an insight into the nature and geologic history of this area.

The Western Highland Rim is the western half of the upland that surrounds the Central Basin of Tennessee. It lies between the basin and the Western Valley of the Tennessee River, and is considered a separate region largely because the political accident of state boundaries cuts off its connections with the Eastern Highland Rim, at least as far as Tennessee is concerned. Looking at the basin as a center and the highland that surrounds it as a whole, ignoring state boundaries, one could speak of a Southern Highland Rim (in Alabama) and a Northern Highland Rim (largely in Kentucky), but these terms have never been generally applied. The Kentucky part of the Rim merges into what is called the Pennyroyal ("Pennyrile") Plateau.

The Western Highland Rim is one of the largest of Tennessee's geologic regions, including something like 7,500 square miles, depending on just where one chooses to set the boundaries; and geologically it is one of the simplest. It is a broad, tilted plateau, standing about 400 feet above the basin to the east and 300 feet above the Western Valley. The shape of the land surface is mostly inherited from the uplift that created the Nashville Dome, from which the Central Basin was eroded in comparatively recent geologic time. That is, the direction and amount of slope of the land surface are roughly the same as the dip imparted to the rock layers by the uplift. The surface configuration of parts of the rim have been modified by recent erosion, particularly in the areas where the Cumberland and Duck rivers cut through, so that the land surface is lower in the northwestern part of the rim than elsewhere. The general elevation of the rim surface, except for the low areas near the rivers, is about 900 feet above sea level near the basin, sloping down to about 700 feet at the edge of the Western Valley.

It is evident that the occurrence of iron ore in the Western Highland Rim but not elsewhere in Middle Tennessee is somehow related to geologic processes that were special to that region. The ore occurs in relatively small deposits, widely scattered over the Western Rim, each deposit perhaps only a few acres or tens of acres in extent and mostly less than 50 feet deep. The amount of ore in one of these deposits might be measured in thousands of tons, or perhaps tens of thousands or even more, but certainly not in millions. The ore mineral was limonite, a brown, heavy, relatively soft material that is made up chemically of iron, oxygen, and water. There are several other iron-bearing minerals, notably hematite, that are mined in other parts of the world and from which iron is more economically extracted. Limonite also characteristically (and unfortunately) contains small amounts of phosphorus and sulfur, which

stay with the iron when it is smelted and impart undesirable characteristics to steel made from it.

The mineral limonite contains almost 60 percent iron, but it seldom occurs in nature unmixed with impurities. In the Western Rim deposits, the ore was liberally mixed with all kinds of useless material (weathered rock, chert, sand, and clay), forcing the miners to separate the ore from a mass that was two-thirds waste. Then, in order to smelt it they required a hot-burning fuel consisting mostly of carbon. This meant charcoal, inasmuch as no local coal was available, and to obtain the charcoal they laid waste to thousands of acres of forest land. An early chronicler of the industry estimated that to keep a furnace with a 12-ton-per-day iron production going for a year required the cutting of 500 acres of forest, and that to keep one going permanently (if the ore held out) would require about 16,000 acres (25 square miles) per furnace, allowing 30 years for timber to grow back before the next cutting. In the year 1873 there were 11 furnaces in blast on the Rim, producing iron at the rate of about 50,000 tons per year. In order for all of these furnaces to operate on a ''permanent'' basis, then, something on the order of 375 square miles of timber would have been necessary to support them.

The use of wood for fuel in those days, both for the iron furnaces and to fire the boilers on the steamboats that plied the Tennessee and Cumberland rivers, accounts for one of the more puzzling place names common to the Western Highland Rim and the Western Valley. Places called ''The Coalings'' appear on numerous maps of the area, leading some people to assume that coal was mined in the region. To the geologist, who knows there is no coal, the name is doubly confusing. Presumably, however, the name arose from the river boatman's habit of referring to any place he took on fuel as a coaling stop, whether the fuel was coal or wood.

By the mid-1870s, concern was being expressed for the future of the iron industry of the Western Highland Rim. Cut over areas were not being replanted systematically, and even natural reforestation was thwarted by the custom of burning over the tracts of cleared land each year to make better pasture. It was widely believed that the day might come when the iron industry would be forced to shut down for lack of fuel. The seers of the day failed to realize that the doom of the industry had already been sealed; that it was already, in fact, in decline.

The iron industry arose on the Western Highland Rim because, obviously, that was where ore was available. But why did nature single out that part of

Eroded piles of waste in these abandoned iron pits near Nunnelly in Hickman County serve as mute reminders of the iron industry that once flourished in parts of Middle and East Tennessee.

Middle Tennessee to sprinkle generously with limonite, while other areas have little or none? The answer to this question is twofold, and both parts are directly related to the geologic history of the region.

In order to have a concentration of a particular substance rich enough to mine one must have, first, an original source for the material, and, second, a natural mechanism for concentrating it into ore bodies. The element iron is quite abundant, making up about five percent of the earth's crust, but in the ore bodies of the Western Rim the concentration runs 20 percent or higher, at least four times its natural abundance.

A phenomenon that should be explained is that the iron deposits all seem to occur at or near the present land surface. Since we know that the present land surface was created by erosion, should we conclude that erosion has everywhere cut down just to the tops of the ore bodies?

Geologists who have studied the ore deposits mostly agree that they were concentrated by the effects of weathering and solution, the same forces that shaped the present land surface. Iron seems to be everywhere on the Western Rim, but only in those places where the bedrock was particularly susceptible to solution and to replacement by iron-bearing solutions was there enough iron concentrated to make ore bodies. This explains the affinity of the ore to the present land surface, but not the absence of ore in other areas that have been subject to the same processes of weathering and solution for about the same length of time. Something else, exclusive to the Western Rim, must also have entered into the process. This exclusive factor must have been the existence of a source of iron, which was later concentrated by these natural processes. The geologic history of the area shows this to be so.

Like the Western Valley of the Tennessee, the Western Highland Rim is underlain by rocks that were formed beneath the sea during a period of time dating back about 600 million years. The oldest exposed rocks, seen only in the deeper stream valleys of the rim and along its edges in the Central Basin and the Western Valley, are only about 500 million years old, but beneath them are thousands of feet of similar but even older rocks, known only through scanty information from a few deep test wells drilled for oil or other purposes, but whose age has been estimated at as much as 600 million years.

The only way to know what happened in a given period of geologic history

Limonite, an iron-bearing mineral found on the Western Highland Rim, is the black "cement" holding together this fractured chert (above). The ore was smelted in charcoal-fired furnaces, such as the Great Western in Stewart County (below), which was used only briefly after its erection in 1854.

is by deduction from the nature of the rocks deposited during the period. When the area of the Western Highland Rim was above sea level, erosion took place, removing some of the rocky record, so that all we can deduce is that the area *was* above sea level. All we can know for certain is that the sea stayed until some time later than 325 million years ago, because rocks of that age are preserved on the rim; and that any younger rocks, equivalent to those preserved to the east and north, were eroded away. Who can ever know what strange creatures roamed what mysterious landscapes where the rim now stands? In other areas, where rocks of the lost time are preserved, weird amphibians, a whole procession of dinosaurs, and the earliest mammals left their bones to the fascinated scrutiny of modern man. But in the rim, all we have is a gap in the record representing about 250 million years and many unanswered questions. The erosion that took place during the gap, in conjunction with the renewed uplift of the Nashville Dome, must have begun shaping the surface of the Western Highland Rim.

But before erosion could proceed far enough to remove the rim rocks altogether, the sea returned. About 75 million years ago the marine invasion that played such a large role in western Tennessee (chapters 1–3) inundated part of the rim as well; and marine sands, clays, and marls were deposited across the older erosion surface. Some of these sediments are still preserved on parts of the Western Highland Rim, and give us our best clue to the origin of the ubiquitous iron ore of the Western Rim. It is not known just how far to the east the last invasion of the sea reached, but because none of the 75-million-year-old sediments are preserved east of the Central Basin it seems improbable that the sea ever reached that far. After the sea achieved its maximum flooding it began a slow, vacillating retreat. West of the rim area it lingered for another 35 million years, but there is no evidence that it ever again reached as far as the rim.

After the retreat of the sea, weathering and erosion became active, and all but scattered remnants of the soft marine sediments are now gone from the rim. But these remnants contain the tell-tale clue—little greenish black specks of glauconite, an iron-containing mineral that is precipitated from sea water. During weathering, it is believed, downward-moving surface water took the iron from this mineral into solution and re-precipitated it in porous layers in the hard rock below, to be left behind when the remainder of the soft sediment was stripped away. According to this theory, little or no iron is found on the Eastern Highland Rim, which is otherwise similar, because the last invasion of the sea never reached that part of Tennessee to deposit such an iron-containing sediment.

What of the future of the Western Highland Rim and its iron deposits? After all, it took nature more than two hundred million years to prepare the ground and fill it with ore bodies like fruit in a pudding, and less than a hundred years for the iron industry to rise, prosper, and die. Is the ore gone, another resource squandered by spendthrift modern man? No, for once we can say that circumstances have forced us into conservation. Much ore remains, perhaps more than has ever been mined. The death of the Western Highland Rim iron industry came about for other reasons. The beginning of the end came in 1844, though nobody knew it at the time, with the discovery of the gigantic, high-grade iron ore deposits of the Lake Superior region. Advancing technology that led to ever increasing substitution of steel for wrought iron accelerated the decline of Tennessee's iron industry. Steelmaking requires huge deposits (whereas Tennessee's are small) and ore low in sulfur and phosphorus (whereas Tennessee's is high). As transportation improved, Tennessee's local industry came into direct competition with the giant steelmakers of the industrial north and could not survive. Ironically, the high-grade ore of the iron ranges of the Lake Superior region are largely exhausted now, and lower-grade ore is used more and more, but still the time for renewed interest in the ores of the Western Rim has not arrived. New processes that will remove the undesirable phosphorus and sulfur must come into use before Iron Hill and Orebank Hollow will again yield up their limonite lodes.

In the meantime, the Western Rim continues, largely, to revert to wilderness (so that Wayne County, not one of the counties of the Unaka Mountain chain, is the most heavily forested in Tennessee), and each census shows the population of this once bustling region to be lower. Andrew Jackson, returning from New Orleans along the Natchez Trace, could not have traversed a country more lonely than parts of the rim today.

5. Bull's-eye Center of Tennessee: The Central Basin

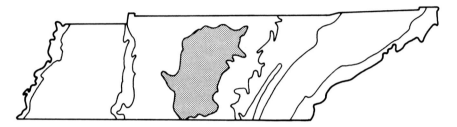

Unraveling the physical history of the Central Basin involves a series of riddles. How can an area be simultaneously a dome and a basin? Why are there fossil remains of sea-dwelling organisms in limestone more than a thousand feet above sea level? Why are some limestones in the area so rich in phosphate? What has become of the thousands of feet of sedimentary rock that once arched over Middle Tennessee and connected the Cumberland Plateau on the east with the coal fields of Western Kentucky and Illinois on the northwest and with Alabama on the south? And how were major rivers able to cut channels directly across that arch (the Nashville Dome, mentioned in earlier chapters), where now lies a basin more than 600 feet deep?

Few cities in this country could claim the title "Dimple of the State" more validly than Nashville, if it chose to appropriate that name. For Tennessee's capital city truly rests in a huge dimple, the Central Basin, one of the physiographic regions of Tennessee. However, Nashville is situated near the northwest edge of the basin, and Murfreesboro, only fifteen miles northeast of the basin's center, might have a better claim to "dimplehood."

A depression carved out of solid limestone that is said to support more bluegrass than all of Kentucky, the oval basin points directly toward the more famous basin around Lexington, Kentucky, that gives the Bluegrass State its nickname. Nearly 50 miles wide, as accurately as a ragged-edged oval can be measured, the basin extends almost to the Kentucky line on the north and just barely into Alabama to the south, as though an effort to disembowel the state with a monstrous shovel were misdirected a little south of center. Since the depth of the basin, about 600 feet, is slight compared to its area of more than 5,000 square miles, it is easy to understand why so few people realize they are down in a hole when they visit Middle Tennessee. Standing at ground level in the center of the basin, it is impossible to see any part of the elevated Highland Rim that completely encircles the basin.

What answers can geology give to the riddles of basins within domes, marine fossils on dry land, limestones so rich in phosphate as to constitute "natural fertilizers," missing rock layers, and rivers that appear to have flowed across hills? The first riddle is easily answered. The Middle Tennessee area is a topographic basin carved by stream erosion from the top of a structural dome—the Nashville Dome. Many millions of years ago an unimaginably powerful force, the origin of which is still not clearly understood, pushed upward from within the earth and bent flat-lying layers of sedimentary rock into a broad, gentle arch of approximately the same dimensions as the present Central Basin. At intervals since that time the uplift has been renewed, and the structure thus imparted to the rocks can still be seen around the periphery of the basin where the rocks slope gently down from the former crest. At the center of the basin the oldest rocks are exposed. Proceeding outward are successively younger and higher layers of rock that form roughly concentric rings around the center point. This bull's-eye pattern is more than a geological curiosity; it helps to account for the topography that now exists.

Uplift of the dome caused two important effects: soft, relatively soluble limestone layers were lifted high above sea level, and both limestone and the overlying harder rocks (like those still preserved on the Highland Rim) were cracked where they were stretched over the dome. As soon as the crest of the dome rose above the sea and the first rain fell upon it, erosion began. Water flowed down the flanks of the dome, collected into streams, and began to cut valleys. The courses of streams were controlled and localized by the cracks; and the highly fractured rock at the crest of the dome, where the gradient was also steeper, was removed faster than less broken and topographically lower rock in areas surrounding the dome. As soon as erosion of the dome had removed the tougher rock, which still caps the rim, a new factor came into play: a difference in hardness and especially solubility between basin rock and rim rocks. Erosion proceeded to scoop out the basin to a depth equal to that of the major streams that were cutting gorges through the harder rim rock to the west and south.

Speculation as to the amount of rock that once covered the top of the dome is fascinating, but perhaps futile. The rock is now gone, and only indirect evidence is available to estimate the maximum height of the former dome. It seems relatively certain that younger rocks still preserved on the Highland Rim once stretched completely across the dome, and almost equally certain that at least some part of the still younger rocks now limited to the Cumberland Plateau also covered this area. An important bit of evidence in this geological

puzzle is supplied by Short Mountain, a former part of the plateau that has been separated by erosion and now stands isolated. Capped by sandstone and conglomerate, Short Mountain looms up near the eastern edge of the basin atop the Highland Rim in Cannon County. The highest point on Short Mountain is 2,074 feet above sea level, higher even than the nearest edge of the plateau, about twenty miles to the southeast. Geologists speculate that if plateau rocks were deposited to the very edge of the basin on the east and are found in great thickness to the northwest of the basin in Western Kentucky, then at least some part must have once lapped across the dome, to be removed later by erosion. If it can be assumed that the entire thickness of younger rocks preserved to the east and west once connected across the dome, then the top of the dome could have towered more than 6,000 feet above sea level. On the other hand, the rate at which the dome was uplifted is not known, nor is the average rate at which erosion reduced its height. The two processes could have been nearly equal; if so, the original dome may have reached little higher than the present Short Mountain.

The uplift of the dome readily explains the matter of limestone beds full of marine fossils now on dry land. Indeed, it gives a rough measure of the amount of uplift, since such fossils were deposited on the sea bed a little below sea level and are now raised several hundred feet above it. And the heavy fossil content of some of the rock layers also provides a clue to the origin of the rich phosphate deposits of southern Middle Tennessee. Some of the fossils, in particular a tiny, coiled, snail-like form called *Cyclora*, are of organisms that were able to extract phosphate from sea water and make their shells from it. Other fossils were the right size, or were broken to the right size by current action to be readily replaced by phosphate in a chemical exchange between sea water rich in dissolved phosphate and the limy material of the original shells. This gave rise to a limestone rich in phosphate, but not rich enough to be mined just as it was. Later, when the phosphatic beds were lifted above sea level and erosion exposed them to weathering near the ground surface, much of the remaining limy material of the phosphatic beds was leached away, leaving a residue doubly enriched, the so-called brown phosphate that is the basis of an important mining industry today.

The geologic history of the Central Basin, like that of our other geologic regions, is relatively brief measured against the age of the earth and the universe. The oldest rocks exposed at the surface date back less than 500

Yearlings graze on a horse farm in the Central Basin of Middle Tennessee, a depression carved out of solid limestone which is said to support more bluegrass than all of Kentucky.

million years. Deep drill holes to about 5,000 feet below the surface, however, go through progressively older sedimentary rocks, the oldest less than
600 million years old. Below the oldest sediments is granite, a type of igneous
rock formed far beneath the earth's surface. The granite has been radiometrically dated at more than 1 billion years, which leaves a time gap of more than
400 million years between it and the overlying sedimentary rocks. From these
facts can be deduced a skeletal outline of the history of the area, beginning
with a 3.5-billion-year blank from the birth of the planet to the formation of
the granite encountered in our deepest drill holes. Because granite is formed
well below ground surface and sedimentary rocks at or near the surface, the
gap of more than 400 million years between the granite and the oldest
sediments must have included at least an episode of erosion that removed
whatever rocks originally lay above the granite, followed by the invasion of
the sea in which the sediments were deposited. It is not known with certainty
when the dome first began its upward thrust, but depositional patterns in
470-million-year-old Ordovician rocks that appear to be influenced by the
shape of the dome suggest that it is at least that old. For more than 300 million
years after the first recorded marine invasion, which began nearly 600 million
years ago, the region was mostly submerged beneath the sea, while thousands
of feet of marine sediments were deposited. During most of this time the dome
was intermittently active, rising occasionally above sea level so that periods
of erosion occurred and parts of the sedimentary deposits were removed,
followed each time by a rise in sea level or subsidence of the dome and
renewed marine deposition. Since about 300 million years ago, however, the
rocks show no evidence that the sea ever again covered the Central Basin. The
marine invasion that covered West Tennessee up until 40 million years ago
may also have reached across the dome; but if so, none of its deposits remain.
Essentially, then, the last 300 million years in Middle Tennessee have been a
time of uplift and erosion, during the latter stages of which—40 million years
or so—the Central Basin of today took form. A piddling time interval, did
someone say? Perhaps so, but enough time to allow over 50,000 years for the
removal of each foot of rock that once arched across the dome, and to spread
the eroded sediment across the shallow sea floor that has only recently been
uplifted to form the Gulf Coastal Plain. Given the span of millions of years
and the immeasurable forces at nature's command, anything is possible, even
the scooping of a 5,000-square-mile basin out of the top of a dome.

*Short Mountain (centered on the horizon) was once a part of the Cumberland
Plateau but now stands isolated, separated from the Plateau by years of
erosion. It looms atop the Eastern Highland Rim, which drops off here to the
Central Basin (foreground).*

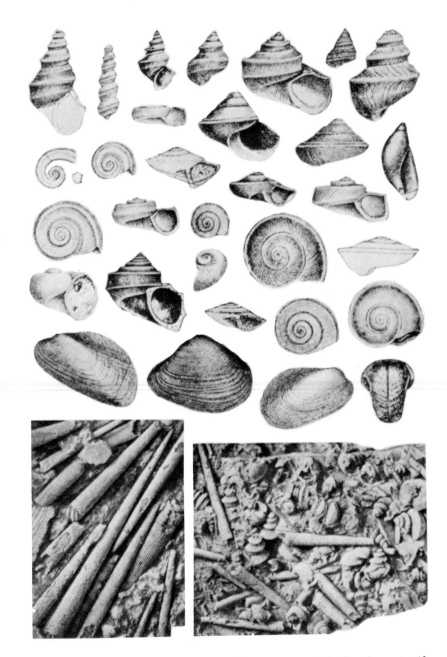

Marine fossils, such as these found in limestone near Murfreesboro, testify that the area, now at several hundred feet elevation, was once below sea level. Oldest exposed formation in the Central Basin, the Murfreesboro limestone crops out again in the Sequatchie Valley uplift.

6. Plateau of the Barrens: The Eastern Highland Rim

The Old Stone Fort, near Manchester on Tennessee's Eastern Highland Rim, is a mysterious structure whose origin has baffled generations of historians, archaeologists, and other curious citizens. Built on a flat-topped point of land between the Duck and Little Duck rivers and almost isolated by bluffs that fall away to the rushing waters below, the walls of the "Fort" are carefully placed to block the few lines of approach where the bluffs can be scaled. It was only natural that the early white settlers of the region, seeing a partly walled enclosure built of heavy stone blocks and earth in a naturally defensible position, would assume that the structure's purpose was defensive. But for whose defense was it built, and against whom?

Speculation over the years has linked the enclosure with beleaguered survivors of de Soto's expedition to the Mississippi; to descendants of Eric the Red, come down across the continent from ancient Vinland; or even to the legendary Madoc, fabled Welsh wanderer of the British Middle Ages. But a 1966 archaeological excavation unearthed charcoal from the builder's fires, with a radiocarbon date of near the second century A.D., well before even Madoc. And so the structure must be credited to the Indians, whose nearest established village of that period was only one mile away.

The origin of the fort (or ceremonial enclosure, as most archaeologists now prefer to think of it) is not its only link in the chain of Tennessee history. Members of the punitive expedition against the Southern Cherokee towns in 1794 camped within its walls on the second night out from Nashville. Later, during the Civil War, a powder mill was built on the Duck River near the Fort, and still later, Union cavalry, in pursuit of retreating Confederates, likewise bivouacked inside.

The Old Stone Fort is interesting enough as a historic site, but there is the added dimension of geologic factors that affected its location, the shape of the country on which it is built, and the materials used in its construction. Some of the rock in the walls is chert from the Fort Payne Formation of the Highland Rim, a material so inhospitable that it gave rise to the treeless, rocky areas called "Barrens" that mystified early travelers. Other lumps are limestone and black shale laboriously carried up from the falls of the Duck and Little Duck rivers. Such falls are common on rivers that cross the western edge of the rim and flow westward into the Central Basin. The falls on the Duck and Little Duck provided power for the old powder mill that helped keep Confederate guns roaring during the Civil War, and caves farther upstream—caves being another distinguishing feature of this part of the state—furnished the niter used in making the gunpowder. About the only item of geological interest in the Eastern Highland Rim that cannot be connected in some way with the Old Stone Fort is the number of small oil fields on the rim farther north, near the Kentucky line.

The Eastern Highland Rim, like its western counterpart, is part of the upland area that surrounds the Central Basin of Tennessee. It is separated from the Western Rim on the north by the notch cut by the Cumberland River as it flows into the state from Kentucky, and on the south by the state boundary that clips off a tiny portion of the basin for the state of Alabama.

The Eastern Highland Rim is an intermediate level between the Cumberland Plateau and the Central Basin. In general, the Eastern Highland Rim is about 1,000 feet above sea level, standing perhaps 300 feet higher than the average level of the eastern part of the basin, and 1,000 feet lower than the Cumberland Plateau to the east.

For the most part the surface of the rim is relatively even, except where notched by streams that flow into the basin on the west, and except where it is surmounted by eroded remnants of the plateau, mostly toward the east. It might be expected that erosion would cause the rim to slope toward the basin, but oddly enough this is not true. The highly resistant Fort Payne Formation, which forms most of the western edge of the rim, dips gently toward the east and southeast. The formation is so difficult to erode that streams commonly flow down its dip, away from the basin, until they reach a stream large enough to have breached the armored layer, at which point they are able to turn away from the plateau looming ahead and flow back toward the low ground of the basin.

Tough, resistant Fort Payne chert (above) helps form the core of walls of the historic Old Stone Fort and accounts largely for Duck River Falls (below) and for the other falls on rivers that cross the Fort Payne formation to flow westward into the Central Basin.

As a whole, the Eastern Highland Rim averages about 25 miles in width, and its total area is about 2,500 square miles, comprising about 5 percent of the state's land area. Escarpments bound the rim on both sides, up to the plateau on the east, and down to the basin on the west. These escarpments have had a dynamic influence on the shape and climate of the rim, as well as on the economic well-being of its people. They hold, in fact, the clue to the ultimate fate of the rim itself. Because erosion is very slowly eating away at the edges of each escarpment, causing them to migrate ever so slowly toward the southeast, they are, in a sense, in a race; and which one wins determines whether the rim either gets wider or else narrows to the vanishing point.

When the first settlers pushed west of the plateau they were heading for fertile bluegrass basins, one to the north and one to the west, in which Lexington, Kentucky, and Nashville now lie. But to reach either of these earthly Edens, the pioneers had to pass through what they regarded as relatively worthless country. Their opinion of the Highland Rim area was perhaps disproportionately colored by the commonly held notion of those days that any ground that could not support trees could not grow crops. Great areas of the rim were without trees amd were essentially prairie land. Little attention was paid to the area at that time, except casually to bestow names like ''Barren County'' in Kentucky, and ''The Barrens'' or ''Plateau of the Barrens'' in Tennessee.

Properly speaking, only the western edge of the Eastern Highland Rim could be termed barren. In this area the surface rock is the Fort Payne Formation. The story is told that a farmer from this area once collected a sample of his ''soil'' and sent it off for analysis, asking what it needed to allow it to bear better crops. The answer came back, ''Everything, except chert.'' The story is apocryphal, of course, but the Fort Payne Formation does consist mostly of chert, and chert is much better for making arrowheads than for producing tillable soil. The poverty of the soil, together with poor drainage caused by the slope of the land surface away from the natural direction of stream flow (toward the basin), probably accounts, in part at least, for the treeless expanses that gave rise to the unflattering name Barrens. Given time enough, forests can grow on such land, and have in many similar areas; but if destroyed by fire, they would take a long time to reestablish themselves.

Chert, which comprises such a large proportion of the Fort Payne Formation, figures prominently in the geology of the Highland Rim. Chert is a very hard, impure, uncrystallized or poorly crystallized form of the mineral quartz, consisting of silicon, oxygen, and various impurities. The dense, even-grained, compact variety of chert called flint used in making Indian arrowheads is familiar to most people. But how many realize that some of our most beautiful precious stones, such as jasper, sardonyx, agate, and carnelian are also varieties of chert? Even the opal is closely akin to chert, although it is formed differently and has water tied into its chemical makeup.

Accounting for the volume of chert in the Fort Payne and other formations of the Highland Rim is difficult, especially in view of the fact that the Fort Payne and equivalent formations extend far beyond the boundaries of the rim, and even of Tennessee, to states as far away as Indiana and Iowa; and almost everywhere the formation is loaded with chert. How does chert originate, anyway?

There are two commonly accepted theories about its origin, and both may have contributed to the Fort Payne cherts. One theory holds that chert is deposited on the bottom of the sea as a gel, derived from volcanic sources or from colloidal silica emptied into the sea by river water. The other theory is that cherty rocks were originally sediments very rich in extremely fine but solid particles of silica derived from the erosion of quartz-rich rocks. Weathering and solution over a long period of time then removed most of the soluble parts of the sediment and then transported and redeposited the silica particles in the form of chert.

Whatever its origin, chert is very hard and almost insoluble in water. For that reason it forms a great barrier to erosion, holding back the wasting away of the Highland Rim escarpment quite as effectively as the much thicker resistant rocks capping the Cumberland Plateau resist erosion of the escarpment to the east.

Chert is also responsible for one of the most scenic aspects of the more westerly escarpment—the numerous and beautiful waterfalls found wherever a stream crosses the boundary between the Fort Payne and the lower, less resistant formations of the basin. Directly beneath the Fort Payne Formation is the Chattanooga Shale, source of the black shale slabs used in the walls of the Old Stone Fort; and beneath that is soft, easily eroded limestone of the Leipers Formation. Water flowing off the hard Fort Payne onto the softer formations below cuts down like a buzz saw to the level of the basin, leaving a lip of Fort Payne over which the water tumbles in cascades and falls. Swirling eddies in the pools below the falls undercut the rock below the lip, creating overhangs that occasionally break away and fall, so that as time goes by the falls slowly migrate upstream.

Not all of the Highland Rim can be classified as Barrens, by any means. The tilted rim of Fort Payne at the edge overlooking the basin forms a protective barrier against erosion, so that to the east higher formations are preserved, nestling against the foot of the plateau and passing beneath it. Some of these formations weather to very fine soil, and in particular the deep-red soil of the

St. Louis Limestone forms some of the best farm land in Middle Tennessee. The topography, also, as one moves east toward the plateau, becomes more rolling, and here and there steep-sided outlying remnants of the plateau rest on the Highland Rim surface, preserving parts of still higher geologic formations.

This eastern part of the rim is one of the nation's best examples of what is called Karst topography, meaning that it is a region of numerous caves and sinkholes, the name deriving from a similar area around Karst, Yugoslavia. The lower slopes of the plateau escarpment and its outlying remnants are mostly limestone of a particularly pure composition, very soluble in slightly acid ground water. Consequently they contain numerous caves, some very large.

Cumberland Caverns, near McMinnville, is reputed to be the second largest cavern in the United States in terms of number of miles of underground passageways, ranking immediately behind Kentucky's Mammoth Cave. Cumberland Caverns was for many years known as Higginbotham Cave; and the origin of that name is an interesting example of the folklore that surrounds the subject of caves in general. Mr. Higginbotham, for whom the cave was named, is said to have become lost inside when his torch went out. It was three days before he was rescued, during which time his hair had turned white from the ordeal. According to the legend, when he finally escaped he promptly traded his cave property for land nearer McMinnville containing, presumably, no caves.

Although caves vary in shape, size, and degree of complexity of the branching passages, most are fairly simple in plan and easy to navigate. Their origin is fascinating, however, being perhaps the nearest approach in nature to the operation of textbook chemistry. Limestone (mostly calcium carbonate) is only slightly soluble in pure water; but the addition of very small amounts of carbon dioxide dissolved from the air produces dilute carbonic acid, which alters the limestone to calcium bicarbonate, which is extremely soluble in water. Acids formed by the decay of organic matter also aid in the solution of limestone. The slightly acid waters seep into the limestone along joints or bedding planes and, over a long period of time, very slowly enlarge them, forming caverns that range in size from mere cracks to extensive cave systems with miles of passageways.

Most of the original work of cave formation is thought to have occurred

Morrill Cave near Bluff City is similar to the many limestone caverns that characterize the eastern part of the Highland Rim. They range in size from tiny openings to extensive cave systems with miles of passageways.

below the water table, that is, in the zone where any open space is filled with water. That most caves existing today contain large, open passages and chambers, sometimes with streams flowing through them, is taken as evidence that erosion has reduced the level of the countryside around these caves, draining the original water from them and leaving open spaces into which new streams can be diverted. The draining of the water also left the cave roofs unsupported by water pressure, so that roof breakdowns can and do occur, enlarging the caves and filling the lower levels with a rubble of broken rock. When the progressive breakdown of a cave's roof reaches upward to the ground surface, a sinkhole is formed.

The later stage in a cave's development, when it is above the water table, is also the time that most of the weirdly beautiful cave formations are created, and when bats leave the guano deposits from which niter is derived. Water seeping downward from the surface takes limestone into solution; then, when the solution emerges into the cave it redeposits the material as pendants hanging from the roof (stalactites), or hornlike stalagmites projecting up from the floor (where water dripped from a stalactite above.) At places where a film of water flows down a wall it leaves draperylike masses (flowstone) coating the walls of the cave. Some of these "cave formations" take bizarre shapes that have led people to apply a variety of fanciful names to them.

Talk of water flowing through dark, subterranean passages leads directly enough to the question; does anything else of value—oil, for instance—similarly flow beneath the surface of the Eastern Highland Rim? Many people visualize oil as occurring in great underground lakes or streams. In this sense the term "oil pool" is a misnomer, because oil actually occurs as a liquid that fills interconnected pore spaces in solid rock. If a well is drilled into the oil-bearing rock, the oil can flow into the drilled opening, but there are no lakes of oil. Sometimes, however, the pore spaces are quite large, on the order of inches, perhaps, and numerous, resulting in rock that is literally honeycombed with oil; and in other instances the oil fills joints or enlarged bedding planes, so that a large volume can flow very rapidly into a drill hole and out to the surface (a "gusher".)

Tennessee has never been a major oil-producing state. In fact, only a few of the considerable number of small producing oil wells would genuinely qualify as "gushers." Two of these, the Jackson well and the Bob's Bar well, were on the Eastern Highland Rim. The Jackson well—located in the Spring Creek oil

Except for having wooden rather than metal towers, the rigs used in drilling the famous Jackson and Bob's Bar oil wells would have closely resembled this structure photographed near Michie in 1946.

field—was drilled in 1866, only seven years after the nation's first oil well was drilled in Pennsylvania. The Jackson well came in with such force that oil gushed 30 feet above the ground. Because there was no way to capture the oil, the operators stood by helplessly for three months while it flowed off down Spring Creek, thousands of barrels going to waste. The remarkable fact about this well is that the oil came from a depth of only fifty-two feet. It is difficult to understand how a confining pressure that blew oil thirty feet in the air initially and continued out of control for three months could be maintained from that negligible depth.

The Bob's Bar well (completed in 1896) was equally frustrating to its backers. The oil here was deeper, 275 feet, but still ridiculously shallow, and it blew out of the hole to a height of from 5 to 20 feet in a solid 6-inch stream. But the oil caught fire immediately, destroying derrick, tools, tank, and all other equipment. After the well ceased to flow, however, and the fire was extinguished, there was still enough oil remaining to produce 600 barrels per day for a while and more than 18,000 barrels eventually.

Many small, shallow oil and gas fields have been located on the Eastern Rim since those early days, mostly in the northern part. Although little or no production comes from them today, a good possibility exists that more will be found, possibly at much greater depths.

The origin of such features as niter-producing caves, oil fields, waterfalls, and even the ubiquitous chert are all intimately connected with the geologic history of the Eastern Highland Rim. The earliest history is very similar to that of the Western Rim, but about 325 million years ago, at the beginning of the Pennsylvanian period, the sea withdrew from the western part of the state and the histories diverged. As in the Western Rim the oldest rocks are the products of ocean sedimentation, mostly limestones and dolomites that began to be deposited almost 600 million years ago, during the Cambrian period, on the crystalline basement complex far below today's rim surface. For at least the next 250 million years this process continued, the rocky record of the millennia stacking up inch by inch into the thousands of feet of sedimentary rock now underlying the rim. Only the top or most recent part of this sedimentary sequence can be seen today, the part that the present erosional cycle has exposed in the escarpments at the edges of the rim. The earlier history here, as elsewhere in the state, is imperfectly known from a few drill holes, a certain amount of geophysical work, and a wealth of informed speculation.

Whereas marine deposition was apparently interrupted on the Western Highland Rim about 325 million years ago, it continued in the area of the Eastern Rim for at least another 40 million years. The Cumberland Plateau, just to the east of the rim, is capped with these younger rocks, as are many of the isolated mountain remnants, such as Short Mountain, that are scattered across the rim. There is little doubt that all of these outliers were once part of the plateau itself, now detached by erosion, and that the Pennsylvanian rocks that cap them once extended as a continuous sheet at least as far west as Short Mountain, which practically hangs over the Central Basin, and possibly all the way across the top of the Nashville Dome to connect with deposits of the same age in the Western Kentucky coal field.

But that is as far as the physical record can be extended in the Eastern Rim. Unlike the Western Rim, there is no record that the later marine invasion, which covered the western part of the state during the interval from 70 to 40 million years ago, ever reached this far. Probably this part of Tennessee has been dry land for more than 230 million years, since the end of the Paleozoic era. Although the physical evidence to show what happened here during the last 230 million years is missing, reasonable interpretations can be made based on the geologic record preserved in neighboring areas to the south and west.

First, if erosion has been going on for 230 million years, it is certain that much rock has been stripped away from the Eastern Highland Rim. The huge gaps that exist between the outliers of the plateau provide ample evidence of this erosion, as do the sediments deposited on the coastal plain to the south and west that bear suspicious resemblances to the rocks of the plateau top. Still more evidence is found in the greater and greater thicknesses of the younger rocks found eastward from the present edge of the plateau—younger layers that probably once extended across the rim as the detached remnants of the lower layers still do.

Another reasonable inference that can be made about this great lost time is that the Nashville Dome and its connections with the Cincinnati Arch and the Ozark Dome must have stood high, at least during the later part of the interval. After all, there must have been *some* barrier that prevented the sea from spreading eastward across the entire state as the earlier seas had done.

Who can foresee the victor in the erosional race between the Highland Rim and plateau escarpments? Surely, one would think, the fact that the plateau escarpment is 1,000 feet high and the rim escarpment only 300 feet, gives a definite advantage to the Central Basin in the battle for supremacy. To push each escarpment a given distance to the southeast, less than one third as much rock need be eroded from the rim as from the plateau. So, it would seem, the rim must eventually disappear, leaving a great escarpment rising all the way from the bottom of the basin to the plateau crest. That would be an impressive sight, but countless generations must live and die before that outcome is proven to be correct.

7. Sky High Table Land: The Cumberland Plateau

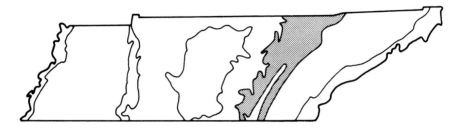

When Daniel Boone went hunting westward beyond Cumberland Gap in the 1760s, he was gone for months. The hunting was good, and he had to walk, but for months? And when James Robertson led his advance guard of settlers to the Cumberland River bluffs in 1779, they sent the women and children, together with the heavy farm equipment, by flatboat down the Holston and Tennessee rivers and on to the Ohio River, up the Ohio to the Cumberland, and up the Cumberland to Nashville. Unlike Boone, the flatboat flotilla was unlikely to go the long way around just to enjoy the scenery. There was a compelling reason for the devious routes to Middle Tennessee and Kentucky. Standing athwart the direct path like the Great Wall of China, but a hundred times higher and 50 miles across, is the Cumberland Plateau, a very real barrier to western migration in the eighteenth century. No roads crossed it, and only a few foot paths like the great Warrior's Path through Cumberland Gap gave access to the interior. Both the Boone and the Robertson parties traveled through Cumberland Gap and across that "dark and bloody ground," Kentucky—Boone to turn north along the Wilderness Road, Robertson south to Middle Tennessee. For many years people preferred to go around rather than across the plateau, and even today the building of roads across the plateau presents unique difficulties.

What kind of feature is this that so inconvenienced our ancestors? The Cumberland Plateau is only a part of the Appalachian Plateau that extends in a southwesterly direction all the way from the southern border of New York to central Alabama, crossing ten states on the way. Its character differs considerably over this great distance, and in fact within Tennessee itself.

Most people become acquainted with the plateau itself by crossing it on I-40 between Nashville and Knoxville or on I-24 at Monteagle. They see a broad, flat-topped ridge one thousand feet higher than the Great Valley of East Tennessee to the east or the Highland Rim to the west. Rimming the plateau edge is an almost continuous line of cliffs, broken by narrow, steep-walled, stream-cut notches running back into the tableland.

The Tennessee portion of the Appalachian Plateau embraces about 4,300 square miles, about one-tenth of the state's area. Along the Kentucky-Virginia line the plateau is about 55 miles wide, but it gradually narrows to about 38 miles near Chattanooga.

The very different appearance of the eastern and western edges of the plateau shows the effect of geology on topography. The eastern edge is an abrupt escarpment, straight to smoothly curving and only slightly notched by drainage that empties eastward into the Tennessee River. The western edge is very ragged and deeply incised by the Cumberland, Duck, and Elk river tributaries that drain it. Why the difference? The answer has to do with early compressional forces, the results of which show up dramatically on the plateau. Hard rock layers were folded like so much spaghetti during the Appalachian mountain building episode 250 million years ago, near the end of the Paleozoic era. These forces, originating somewhere east of the Appalachian Mountain chain, reached far enough to the west to bend the eastern edge of the plateau, but not the western. All along the eastern edge the rocks were folded or broken (or both), so that the same rock layers that form the flat-lying rim-rock to the west are tilted in the east, in some places even standing vertically in towering crags and pinnacles. Where the escarpment is thus armored with sloping sandstone layers, erosion is slowed, and the shape of the escarpment is controlled by the direction of the folds. This factor is almost completely absent from the western part of the plateau.

Mountain-building forces are also responsible, indirectly, for topographic differences within the plateau. Tight folding was largely restricted to the eastern edge of the plateau, but elsewhere the rock was compressed to the point of breaking—which it did. Large-scale breaks reduced the stresses, and great masses of rock moved along these breaks—properly called faults—for as much as 10 miles. Faulting is much more important in the Great Valley of East Tennessee than on the plateau, and will be taken up again in the next chapter; but much of the top surface of the plateau has, in fact, been thrust upward and to the northwest along such faults. The entire group of faults that together form the boundary of the displaced part of the plateau is called the Cumberland Plateau Overthrust fault. It is not a single, simple break, but a complex, interwoven system of faults along which some parts of the plateau moved very little, but others moved for great distances.

The system begins near Elverton, in Roane County, where the Little Emory

River cuts through Walden Ridge. Along the northeastern margin of the huge area that moved, the fault at the edge is like a tear in a sheet of paper, the rocks on one side having been shoved northwestward past the rocks on the other. This fault generally follows the course of the Emory and Little Emory rivers to a point near Catoosa in Morgan County, where the fault system turns a corner and angles off toward the southwest. From there on the fault is a thrust, a gently dipping fault plane along which rocks above the fault have moved up and over other rocks. Characteristically such faults bring lower (older) rocks above higher (younger) rocks, and in this instance older rocks from the southeast are thrust over younger rocks to the northwest. Apparently the amount of movement on the fault decreases steadily toward the southwest, from a maximum of perhaps seven or eight miles near the Emory River to almost nothing at the point where the fault reaches the western edge of the plateau near Spencer. The over-riding mass of rock, then, must have pivoted toward the left as it moved.

Sequatchie Valley, the beautiful, almost ruler-straight chasm that bisects the southern half of the plateau in Tennessee, also owes its origin to faulting and folding. First came the fault, which is one of the subsidiary breaks in the Cumberland Plateau Overthrust system. Rock from the southeast was pushed up and over rock to the northwest along a 180-mile break on the west side of Sequatchie Valley. Movement totaled thousands of feet, and the enormous over-riding block was folded into an arch or anticline. At the north end of Sequatchie Valley the arch is still topographically high (Crab Orchard Mountains), but over most of its length it was so fractured and jointed by the bending that erosion has found it easy prey and scooped it out into a long, linear valley.

Sequatchie Valley and the Crab Orchard Mountains form a convenient line for subdividing the plateau. That part of the plateau west of Sequatchie Valley is called by the name commonly applied to the whole, the Cumberland Plateau. That part of the plateau east of Sequatchie Valley is called Walden Ridge, named for Elijah Walden, one of the famous "Long Hunters" of the Daniel Boone era. Two other subdivisions of the plateau in Tennessee also owe their distinctive character, at least in part, to Appalachian mountain-building. These are the Cumberland Mountains and the neighboring Cumberland Block.

Cumberland Gap (above), seen here from the Tennessee side, afforded a welcome access to the interior for pioneers facing the forbidding Cumberlands. Below, the unbelievable force of early mountain-building stood rock layers on end, as seen where I-75 cuts through the Pine Mountain block.

To visualize the relationship of the plateau to the Cumberland Mountains, think of a table on one end of which is a pile of books. The table is the plateau, and the table top is the resistant cap rock. In the northeastern part of the plateau the cap rock slopes gently downward and disappears below the land surface in a broad, down-warped area centering on the common corners of Morgan, Scott, Anderson, and Campbell counties, reappearing on the far side. The downwarped area is thus completely surrounded by the resistant cap rock that protects it, like a stockade, from the encircling forces of erosion. Thus protected, within the basin stand towering mountains carved from successive layers of flat-lying sedimentary rocks that have long since been eroded away from other parts of the plateau. These mountains, the book pile of the analogy, are mostly shale inter-layered with numerous coal beds, making this the most important coal-mining area of the state. The tallest of these mountains, 3,534 foot Cross Mountain, is the highest point between the Smokies and the Black Hills.

The high ridges of the Cumberland Mountains present a startling contrast to the way the country looked when the coal beds were first deposited. Microscopic examination shows that coal is made up almost entirely of carbonized plant fragments, from plants that grew in ancient swamps much like Georgia's Okefenokee Swamp of today. Extensive swamps of this kind are found only in very flat areas, such as coastal plains or wide river flood plains, not far above sea level. In these places the thick vegetation lives and dies, to fall and sink beneath the murky water, which protects the plant material from rapid decomposition. Succeeding generations of swamp plants grow atop the old, falling in their turn to add to the accumulating thickness of plant material. In time the weight of the accumulating mass squeezes much of the liquid and gaseous constituents out of the lower layers, converting them to a spongy material called peat. Burial of the swamp itself under layers of other sediments adds more weight and compresses the material still further, and deep burial adds the factor of increasing heat to the process of change. All of these factors combine to drive out the liquid and volatile constituents, leaving a thinner and thinner layer that is richer and richer in carbon but poorer in water and hydrocarbons. The material thus passes by successive stages from wood to peat to lignite to coal. The presence of numerous coal beds in the Cumberland Mountains thus tells us a story of many thousands of years of quiet

Viewed from the Catoosa game preserve, the Cumberland Plateau (above) seems to rise toward the distant Cumberland Mountains. Beautiful Sequatchie Valley (below), which was eroded from an enormous but fractured arch, is seen here from Raven Rock near Pikeville.

accumulation of plant material, followed by deep burial, uplift, and erosion to form the mountains of today.

The Cumberland Block is a classic example of topography closely reflecting geologic structure. Straddling the Tennessee, Kentucky, and Virginia lines, this unusual rectangular feature is about 125 miles long by 15 miles wide (only the southwesternmost 35 miles or so are in Tennessee). Approaching the Cumberland Block on I-75 from Knoxville, the most impressive feature is the offset in the chain of mountains west of Lake City. It looks like one end of the mountain mass has been broken off and shoved 10 miles to the northwest, and in fact this is exactly what *has* happened. On three sides—northeast, northwest, and southwest—the block is bounded by faults; and the whole mass has been shoved along these breaks and simultaneously partly rotated, so that the Tennessee end has moved about 10 miles, but the Virginia end, 125 miles away, has moved only about 4 miles. During the movement the overriding mass of rock was folded so that the rocks on Pine Mountain, which forms the leading edge, dip southeast into the block, whereas rocks on Cumberland Mountain dip northwest into the block. Since crowding along the faults at each end of the block has also crimped the rocks into folds that dip toward the interior, what remains is like an enormously elongated, rectangular cake pan, again with mountainous masses of younger, softer rock within the basin protected from erosion by enclosing rims of harder rock.

There are only a few items of plateau nomenclature that have not been discussed. On the east, Lookout Mountain, Lone Mountain, and Powell Mountain are all parts of the plateau detached by erosion from the main mass. To the south, Raccoon Mountain and Sand Mountain are extensions of the plateau running south into Alabama. On the west the remnants left behind by a retreating western escarpment are "too numerous to mention."

The topography of the plateau also exhibits many interesting features smaller in scale than the ones that differentiate regions. Most are caused by the differing resistance to erosion of various kinds of rock. The sides of the plateau itself, the escarpments, are good examples. The vertical bluffs are formed from hard, resistant sandstone that also makes a flat top for the plateau. Below, underlying the gentler lower slopes, are shales and limestones. At almost any place you approach the plateau, if you can get a glimpse of the profile of a slope on some spur, you will see that it forms a sort of graph

Natural bridge at Pickett State Park was carved out of conglomeratic sandstone by the action of running water. The resistant sandstone of the arch provides the caprock over most of the northwestern part of the Cumberland Plateau in Tennessee.

of the hardness of the rocks beneath the slope; the steeper the slope, the harder the rock. The profile is so characteristic that photographs of different promontories, if taken from the same angle and distance, can be superimposed and only minor differences noted.

Because of the kinds of rock over which the streams flow, there are probably more waterfalls on the plateau than any area of similar size in the state. Above the sandstone that forms the rim of the plateau are successively higher layers of sandstone or conglomerate separated by layers of softer, more easily eroded shale. Wherever a stream flows over the edge of a sandstone layer and digs deep into the underlying shale, a pool is scooped out. Eddies and currents set up in the pool by the swift-flowing water undercut the edge of the sandstone layer, which breaks off into a vertical face. The pool is now a plunge basin, with a waterfall on one side. This process is especially characteristic of plateau streams, many of which have waterfall after waterfall along their courses as the stream cuts through successive layers of sandstone, each with its accompanying plunge basin. These are, incidentally, very fine natural "swimming holes." Some of the waterfalls on the plateau are famous scenic attractions, in particular Fall Creek Falls in Van Buren County. At 256 feet in height it is reputed to be the highest falls in the United States east of the Rocky Mountains, and more than twice as high as Niagara Falls.

The character of the Cumberland Plateau has changed surprisingly little since the days of the Long Hunters, despite the multitudinous activities of man. The timber has been cut again and again, but continues, under better modern management, to grow back. Roads now crisscross its surface, but make only a slight impression on its vastness. In some areas coal-mining has scarred the land, but even this ultimate devil in the pantheon of modern environmentalists can eventually be brought under control and the scars healed, if man and nature work together toward that end. Yet still, as of old, the plateau manages to slow or stop man's restless wandering. Occasionally, as at Rockwood Mountain or at Jellico, the mountain shrugs its shoulder and another of our expensive superhighways goes sliding down the slope.

8. Land of the Rootless Ridges: East Tennessee's Great Valley

On crisp, clear autumn afternoons the trees around Knoxville do their bit to add to the "Big-orange" frenzy that sweeps East Tennessee every fall. On "The Hill" thousands of University of Tennessee students loiter in the sun, enjoying the weather and the afternoon respite from classes and labs, while other tens of thousands jam Neyland Stadium to watch the football Vols do battle. Yet, of all these thousands treading the "secure" earth of Knox County, not one in a hundred realizes that this seemingly permanent land has moved all the way from the middle of Sevier County to the place it now occupies.

Of course, Knox County is not the only area that has moved. Lookout Mountain, just west of Chattanooga, has moved thousands of feet to the west, whereas a part of the city itself (the eastern part) has moved several miles. Cleveland once moved about as far as Knoxville did, perhaps fifteen miles. Gatlinburg has moved at least twenty-five miles to the northwest, and the high spine of the Smokies has traveled even farther, some say forty to fifty miles.

There is, of course, a rational explanation for this restless migration of the solid earth, all of it increasing in magnitude toward the southeast but apparently headed in the opposite direction. The answer lies in geologic faults, the same phenomenon that caused much of the top of the Cumberland Plateau to move to the northwest. These breaks in the rock, along which movement has occurred, can have any orientation, from vertical to horizontal, and the movement can be in any direction, up, down, or sideways. The break can be straight or curved, or straight in some areas and curved in others, and the amount and direction of movement can vary tremendously from place to place, or even reverse entirely. The distinctive point about faults is that movement has taken place. Very commonly in looking at rocks, for example

in a road cut, one can see obvious breaks, and the temptation is strong to say "Aha! A fault!" But this is not necessarily true. If you look closely and see that the layers of rock on each side of the break align with each other exactly, without any offset, then what you are seeing is not a fault but a joint (by definition, a break in the rock along which no appreciable movement has taken place). The difference between a joint and a fault is analogous to that between a simple fracture (no offset) and a compound fracture in broken bones.

Much of the movement of the earth in the Great Valley is explained by the fact that the area is literally crowded with faults. Many of the faults are tremendous in size and can be traced along the ground surface for up to hundreds of miles. The amount of movement may also be great, and may, in fact, be measured in miles. Yet by no means did faulting cause all of the movement of the wandering rocks of East Tennessee. Even a short drive across the eastern part of the state will show the traveler many examples of layers of rock that have been folded and twisted without, apparently, having been broken. If one could somehow straighten out all the compressed and folded layers of rock and stretch them out again into the flat layers that were originally deposited, it would become apparent that folding has also contributed its part to the northwestward movement.

There are several kinds of faults (not to mention folds) in East Tennessee, but the most important is the thrust fault, described in chapter 7. Some of these faults extend all the way across the state from Virginia to Georgia and far beyond in both directions.

In East Tennessee most of the thrust faults dip gently to the southeast except where they break up steeply through layers of hard rock from a lower to a higher level. It is generally believed that all of these faults come together in a single major break far below the surface somewhere to the southeast in the lowest zone of soft rock above the hard granites or metamorphic rocks that lie at great depth below the valley. This "master" fault is a theoretical concept, which, if it actually exists, is too deep to have been penetrated by any drilling yet done. It is fact, however, that toward the southeast older and older rocks are found along the faults, showing that they stairstep downward to lower and lower layers. In parts of the Smoky Mountains, in fact, the faults have apparently even cut into the ancient crystalline rocks that underlie the sedi-

Road cuts offer excellent opportunities to examine many of Tennessee's geologic features. The diagonal break in these tightly folded rock layers along I-75 north of Ooltewah, Bradley County, is clearly a fault.

ments, because granites and metamorphic rocks have been thrust over the much younger sedimentary rocks.

The geology of East Tennessee is so complex that it demands explanation and simplification. The Great Valley of East Tennessee is, on the average, about forty-five miles wide, but if all the folds could be ironed out and all the faults pulled back to their original positions before movement, the average width of the valley would be more like seventy miles. And if the Smoky Mountains, which bury the eastern edge of the valley, could somehow be shoved back to where they came from, the width of the valley would be considerably over one hundred miles. What has caused this part of the earth's crust to be compressed in this fashion? One is reminded of the statement that Switzerland would be the largest country in Europe if all of its mountains could be flattened out.

Explanations for the crumpled state of the rocks of East Tennessee take one to the frontiers of theoretical geology. Early thinkers could come up with nothing more satisfying than to say that some great force, of unknown origin, pushed from the southeast and shoved, folded, and thrust the rocks toward the northwest like a bulldozer cleaning dirt off the bedrock at a building site. This

Four principal kinds of faults found in Tennessee.

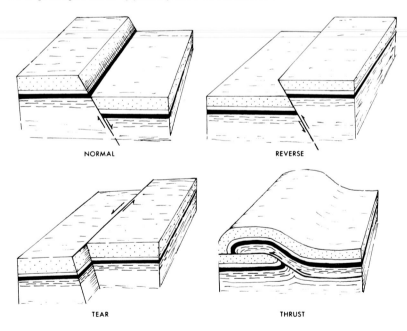

seemed to fit the facts, as far as they were known at that time, but left unanswered the question of what the force was and how it was applied.

Probably the most intellectually satisfying explanation is that the folding and faulting result from sea-floor spreading. Sea-floor spreading is the latest major discovery in the field of geology, and one that has revived and made respectable the old, long-scoffed-at idea of continental drift. Sea-floor spreading results from the upwelling of new crustal rock along a system of fractures in the ocean floor, pushing aside the older crustal plates and literally spreading out or widening the bottom of the ocean. The continents have moved also, carried along by the movement of the sea floor. It is now known that the continents have drifted back and forth several times through geologic time, and a collision between North America and Africa about 250 million years ago may have supplied the compressive force so evident in the deformed rocks of East Tennessee.

However one tries to explain it, the fact remains that the Great Valley of East Tennessee has been deformed in a rather peculiar way, so that the entire area has been telescoped. To the northwest, on the Cumberland Plateau, the amount of horizontal movement caused by the folding and faulting is very little, probably zero at the west edge of the plateau. But the farther southeast one goes across the valley, the more movement there has been, until one arrives at the Unakas, where movement has totaled some tens of miles. To the east of the Unakas in the Blue Ridge and Piedmont the rocks are so tremendously mangled that the amount of movement, though obviously great, is extremely difficult to estimate. Then, beyond the Piedmont the deformed rocks disappear beneath the younger sediments of the Atlantic Coastal Plain, and we are left completely in the dark about what may have happened to the older rocks.

To show how it is possible for the amount of movement to vary systematically in a particular direction, we might compare the thick layer of deformed sedimentary rocks in the valley to a carpet laid down on a slick floor and tacked down at one edge only. A slow, steady push on the other side of the carpet would slide it across the floor, wrinkling it into successive folds until it had finally been compressed as much as the amount of force applied would allow. The front edge of the carpet, being tacked down, would not have moved at all, while each successive fold going toward the other side of the rug would have moved farther and farther.

Since East Tennessee has been so active, geologically, it is easy to understand why geologists find it a fascinating area in which to work. In fact, the Great Valley of the Appalachian region is a classic area for geologic study and has been visited and studied by geologists from all over the world from the

earliest days of the science. The structural complexity would be reason enough for this attraction, but East Tennessee also has another claim to geologic fame. Less spectacular, but just as important in its own way, is the enormous thickness of sedimentary rocks preserved here. Nature, as if wishing to display her wares to best advantage, has folded and sliced up the rocks and arranged them like cold cuts on a hero sandwich so that we can see all of the enormous thickness of rocks, not just the few hundred feet that erosion might expose at a given place where rock layers are flat and undisturbed.

In describing the rocks, piecing together the sequence, and collecting and classifying the fossils, geologists have discovered that not only are the rocks more mangled structurally here than elsewhere, but also that they are much thicker. The sedimentary rocks that bury the crystalline "basement" rocks in East Tennessee are several times thicker than those, for example, in Middle Tennessee. Given the history of squeezing and distortion that has taken place here, it would be expected that the section would be thickened to the same extent that the width of the original area had been narrowed, that is, by 30 to 40 percent. This is not nearly enough to account for the fact that the depth to basement in East Tennessee is on the order of thirty to forty thousand feet, but only 5 thousand feet at Nashville. Something more than structural squeezing must be operative to account for the great thickness of sedimentary rock in East Tennessee. Geologists believe the sediments accumulated in a long, relatively narrow trough occupied by an arm of the sea, and that the sediment was derived from the erosion of nearby land masses and from organic remains. This sounds little different from a bay or lagoon, but there was an additional characteristic that made the difference—as fast as sediment was dumped in, filling the trough, the bottom subsided, leaving room for greater and greater thicknesses of mud, sand, gravel, limestone, or whatever. Something like this must have happened in East Tennessee, because most of the tens of thousands of feet of sediment bear unmistakable evidences, such as mud cracks, ripple marks, and shallow-water fossils, of sediment deposited in shallow water. To have such great thicknesses of shallow-water deposits, the bottom must have sagged at about the same rate as filling took place. Geologists coined the word geosyncline ("earth downwarp") for the sagging depositional trough in East Tennessee, and the term has since been applied to similar features all over the world.

Tremendous compressive forces in East Tennessee deformed these rock layers into a downward fold or syncline (above), seen along State Route 40 near Parksville, Polk County, and into an arch or anticline (below) along State Route 9 in the Douglas Reservoir of Jefferson County.

East Tennessee owes at least one industry—marble quarrying—directly to the continuous downwarping of the geosyncline. Tennessee has ranked second or third in the nation for years in marble production, despite the handicap of having no marble at all, technically speaking. The Tennessee marble is actually a coarse-grained, highly colored limestone that takes a beautiful polish and thus conforms to the commercial definition of marble. It is not, however, a carbonate rock that has been recrystallized by heat and pressure (metamorphism) and thus does not conform to the technical definition of marble. The coarse grains in the Tennessee marble are mostly original, resulting from the grinding up of fossils to coarse fragments by the action of currents in the shallow water of the geosyncline. The subsiding bottom of the geosyncline kept the water just deep enough to maintain wave or current action while banks of the ground up shell fragments accumulated to great thicknesses locally. When this material was colored and cemented together after burial, it formed the masses of "marble" in the Holston Formation of the Knoxville area, and made possible the industry.

Putting the idea of the geosyncline together with the sea-floor-spreading theory, one could visualize the geosyncline as a zone of weakness caused by movement of the sea floor and then depressed by the weight of the sediment dumped into it. The East Tennessee geosynclinal trough was apparently active in this way for many millions of years, until the continents came together again, and the thick body of sediment that had accumulated was deformed into the complex structures seen today.

A similar process appears to be taking place now along the Gulf Coast of the United States. Sediments dumped by the Mississippi and other rivers have accumulated to a thickness of many thousands of feet in a down-sagging trough paralleling the coast, and all that is needed to complete the geosynclinal sequence is for Yucatan to come floating northward to squeeze the sediments like a trash-masher and create a new mountain range. It sounds far-fetched, but not more so than what has apparently happened to East Tennessee.

What kind of countryside might one expect to develop from the effect of wind, rain, and running water on a mass of rocks formed and deformed in the manner just described? Obviously, it would not be flat, neither the rock layers themselves nor the topography eroded from them. To return to our wrinkled rug analogy, there is no way you could push one edge of the rug half way

Rock formations are many times thicker in East Tennessee than in other parts of the state, as shown by the massive thickness of rock in this marble quarry near Knoxville. Note also how the layers tilt.

across the floor and expect it to lie smooth. Throughout the western two thirds of Tennessee the rock layers are flat or nearly so, with a few local exceptions, but not in the Great Valley. There beds of rock standing vertically on end are quite common, and the few limited areas where rock layers are nearly horizontal are eye-catching exceptions to the general rule of steep dips.

Because the many thousands of feet of sedimentary rocks that have been so compressed, folded, and faulted in East Tennessee are also very diverse in their composition, there are great variations in their hardness and solubility, and consequently in their resistance to erosion.

The force that deformed the rocks appears to have come from the southeast, and the structures in the rocks are almost all oriented in a northeast-southwest direction, at right angles to the stress direction. This means that the dipping layers of rock also line up in a northeast-southwest direction, parallel to one another like wind-driven waves approaching a coast. Erosion, in the process of scooping out the Great Valley, has been able to cut deeply into the softer or more soluble layers, but had a much tougher time dealing with the harder rocks and left them standing up as ridges. The sight of one of these ridges, with its rock layers plunging straight down into the earth, gives the impression of great stability, of roots reaching half-way to China to anchor the crust against any possible lateral movement. This is deceptive, however, since almost all of the ridges are underlain at no great depth by fault planes, along which movement has occurred. In this sense these are truly rootless ridges.

The land surface in the Great Valley, for all these reasons, has developed a strikingly orderly arrangement of long, parallel hard-rock ridges separated by soft-rock valleys. Some of these ridges and valleys are extremely long, extending all the way across the state from Virginia to Georgia. Others are not nearly so long, but all are oriented in more or less the same direction. The other name for the Great Valley of East Tennessee—the Valley and Ridge Province—is a reference to this unusual, almost geometrical regularity of the land surface.

The Great Valley of East Tennessee, then, is a relatively low-lying area between the Cumberland Plateau on the west and the Unaka Mountains on the east, and extending diagonally across eastern Tennessee from Virginia to Georgia. The ridges and valleys that seam its length come in all widths, depths, and heights, from a high elevation of 2,620 feet on Clinch Mountain to a low of 640 feet where the Tennessee River exits the valley near Chattanooga. The area of the Great Valley is about 9,000 square miles, about one-fifth of the area of the state.

For so much to have happened in East Tennessee it must have been a very busy place, geologically, for a very long time. Geologic information regard-

ing its history goes back further than that of any other geologic region in
Tennessee, except the Unaka Mountains. The kinds of events and processes
that have occurred here have already been described in general terms, but
perhaps a brief historical summary might bring things into a little clearer
focus, particularly in regard to the more recent shaping of the area.

For more than one billion years there appears to have existed a sagging
depositional trough, being filled and refilled with sediments. Probably at first
this trough was far to the east, or at least the oldest rocks are found in that
direction, but later on areas farther to the northwest also sagged and were
filled with sediments, this process continuing until about 250 million years
ago, late in the Paleozoic era. Then the sea retreated from East Tennessee for
the last time, and mountain-building—probably caused by the collision
between the North American and African continents—folded, faulted, and
altered the rocks, pushing them to the northwest along a series of thrust faults
that stacked up the rock units like shingles and possibly raised up highlands of
a magnitude comparable to the Himalayas today. Then, about 150 million
years ago, during the Jurassic period, the continents began to pull apart again,
relieving the compressional stress and allowing erosion to cut away the
highlands. Since that time there have been at least two cycles of erosion,
reducing the level of the area almost to sea level, followed by renewed, but
this time vertical, uplift. The region is probably now in the middle of a third
cycle of erosion, and the ridges and valleys of East Tennessee could therefore
be regarded as third generation descendants of the original high, mountainous
mass built by that long-ago collision of continents.

The word *fault*, in this day and time, is an anathema. Mention that the Great
Valley of East Tennessee is shot through with faults, and people immediately
think of the San Andreas fault and the San Francisco earthquake. The truth,
though, is that none of the East Tennessee faults has been active for millions
of years. If this were not enough to allay anxiety, then an examination of one
of the faults should do the trick. Because good outcrops are hard to find it is
difficult to examine the faults, but those that are exposed seem to have been
completely sealed with mineral deposits and finely ground rock flour. In many
ways they seem to be like broken bones that, having once healed, are forever
afterward stronger at the point where they were broken than anywhere else. It
seems probable, in fact, that if stresses were somehow renewed in the Great
Valley the breaks would probably occur at entirely new locations.

This is not to say that movements, with consequent earthquake waves, do
not occasionally take place in the Great Valley. But such movements are at
considerable depth, down in the zone where rocks are somewhat plastic, and
the pattern of movement bears no discernible relationship to the ancient faults

that are such prominent geologic features of the area. And, as earthquakes go, they seem to be relatively minor; so that although the possibility of feeling an earthquake tremor some time or other is pretty good, the probability of major damage is slight.

If anything is evident from this discussion of the geology of the Great Valley, it should be the fact that it is complicated. In trying to determine what rock units at one place are the same as what units some place else, the geologist studies the physical character of the rock, in an effort to determine what kind of depositional process or environment the rock represents. He can then deduce from this the kind of pattern that rock type might have, by comparing it with modern examples of the same kind of deposition, and he can logically assume that rock outcrops not too far apart and of the same general kind probably are connected beneath the soil cover that obscures the rock layers. None of this can be done very readily in the Great Valley, because just when he thinks he has figured out what environment a given unit belongs to, and one more exposure would prove it, he runs up against a fault and on the other side is some completely different rock type, probably of a completely different age. Or he finds an exposure on one side of a valley, and on the other side more rock very similar to it, and assumes them to be the same, only to discover later that a fault goes down the middle of the valley and that the two rock units actually were originally deposited many miles apart. All this in an area where, instead of the usual few hundred or few thousand feet of rock layers the geologist must learn to recognize, there are tens of thousands of feet. The sheer volume of rock to be studied would be problem enough for anybody, but on top of that add in the faults, folds, and other structural complexities, and one reaches a level of difficulty so great that in some unusually complicated and poorly exposed areas the geology is literally indecipherable. This might be considered just one of those things nobody worries about except a few people involved in scientific research, but for the fact that an understanding of the basic geology is extremely important in the search for mineral deposits, such as the zinc, iron ore, barite, and even marble that have contributed so much to the wealth of this part of our state.

All in all, the difficulties a geologist must contend with in the Great Valley are enormous, but the geologist does have one thing going for him. As far as we know, Knoxville and Gatlinburg are not getting any closer together these days, and we could not always say that.

9. Tennessee's Eastern Rampart: The Unaka Range

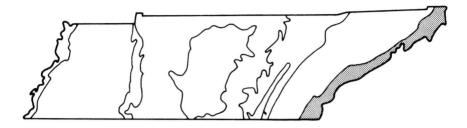

The time is the first half of the nineteenth century. Overhead a hot summer sun bears down on blue mountain ridges, but in the shadowed gorges men dressed in frontier garb squat ankle-deep in icy torrents, skillfully swirling water and gravel from shallow pans until nothing remains but black sand, flecked with the hypnotic gleam of pure gold. Forty-niners, one might guess, at Sutters Mill in California. But not so! The year is 1831, and the place Coker Creek in Monroe County, Tennessee, at the start of Tennessee's almost forgotten gold rush. For nearly a hundred years the gravel bars of Coker Creek were worked and reworked for an ever diminishing golden reward. Even though the area never made it as a major gold district (total production was less than a quarter of a million dollars in gold) a knowledgeable (and energetic) gold-panner can still find enough "color" in his pan to see.

The Coker Creek district is only one of a number of gold fields scattered up and down the Appalachian Mountain chain, and by no means the most important. Still, the old cliché, "there's gold in them-thar hills" applies.

What mountains are these where gold can be plucked from the streams? There is a plethora of names by which they are called, depending on where you are. The high ridges around Coker Creek are called the Unicoi Mountains, and farther north such names as the Great Smoky Mountains, Unaka Mountains, Stone Mountains, and Iron Mountains apply to various ranges. Some geographers, in a laudable effort to simplify terminology, refer to the entire mountain mass from Maryland to Georgia as the Blue Ridge. Most of them, however, prefer to restrict this name to the easternmost ridge rising above the Piedmont of Maryland, Virginia, and North Carolina, well to the east of the Tennessee line. Early geologists called all of the Tennessee part of the mountains that separate the Volunteer State from North Carolina the Unaka Range, and for simplicity this usage will be followed here.

The diversity of names correctly suggests a considerable diversity in the mountains themselves. The mountain mass is widest in the central part, near Gatlinburg, where the name Great Smoky Mountains applies, and narrows both to the north and south. The widest part is also the highest, with numerous heavily forested, gently rounded swells over 6,000 feet in altitude. The highest point in the whole Appalachian chain, Mount Mitchell (6,684 feet above sea level), is in this area, although some miles east of the Tennessee line in North Carolina. Clingman's Dome, the highest point in Tennessee, is not too far behind (or below) at 6,642 feet. To the north, in Virginia, the range narrows drastically from its seventy-five mile maximum to about fifteen miles where it joins the Blue Ridge, and diminishes in height to a relatively insignificant 3,000 feet or less. To the south the dwindling is less drastic, and the range is still forty-five miles wide with peaks up to 4,000 feet high where it marches into Georgia.

To further emphasize the diversity of the Unaka Range, its character changes from west to east as well as from north to south and bottom to top. Along the western edge of the mountains the individual ridges (Chilhowee Mountain, for example) are long, narrow, and knife-edged, aligned parallel to the trend of the range and similar to the lower, more subdued ridges of the Great Valley of East Tennessee. The farther east one goes, however, the less parallelism of ridges there is, and in the main mountain mass along the state line the tumbled confusion of peaks and valleys seems to have no pattern at all. An aerial view, however, or a good topographic map shows a system where the eye of the earthbound observer sees none. The distribution of the ridges and valleys is closely controlled by the major streams that drain the area, the French Broad, Pigeon, Little Tennessee, Hiwassee, Tellico, and Ocoee rivers.

In the core of the mountains each stream, as one proceeds upstream, divides symmetrically at the junction with each tributary, so that in viewing it from above the drainage pattern resembles a tree, with the main stream as trunk and successively smaller tributaries the branches, limbs, and twigs. The mountain ridges separate the tributaries in this tree-like drainage pattern, with the highest ridges of all occurring at the major divides that separate one watershed from another. In the core of the mountains, where the pattern is difficult to see, it is the streams that seem to control the distribution of the ridges, while to the west, where the topography seems better organized and more predictable, the

Long, linear ridges and valleys of upper East Tennessee, seen across the fire tower on Chimney Top Mountain (above), contrast sharply with the confusion of peaks (below)—Charlie's Bunion in the foreground—in the high backbone country of the Smokies.

trend of the ridges controls the location of the streams. To the west the streams closely parallel the ridges except for short stretches where the streams break through the ridges. This pattern of short segments cutting through the ridges to connect much longer segments that parallel the ridges resembles an old-fashioned trellis, and indeed is commonly called trellis drainage. The eastward transition from linear ridges and trellis drainage to irregularly forking ridges and a treelike drainage pattern is caused by differences in the nature of the rocks that make up the mountains.

One more point should be made about the drainage system of the Unaka Range. Almost all of the Unaka Range drains northwestward, into streams that flow by way of the Tennessee, Ohio, and Mississippi rivers into the Gulf of Mexico. This seems anomalous, since the highest peaks of the Appalachians lie between the headwaters of these streams, and it looks like water flowing down their eastern slopes would flow toward the Atlantic rather than the Gulf of Mexico. In fact, however, the divide between the Gulf and Atlantic drainage lies miles to the east, at the crest of the much lower Blue Ridge, and that part of North Carolina west of the Blue Ridge drains westward through the highest mountains east of the Rockies.

Some of the best known topographic features of the Unaka Mountains are not mountains at all but are strange, flat-bottomed, mountain-circled valleys called coves. The most visited of these probably are the three that lie within the Smoky Mountain part of the range; Cades Cove, Wear Cove, and Tuckaleechee Cove. Although different in shape and size, all of these coves have geologic features in common that point to a similar origin. And all of them seem to be out of place, to interrupt the various progressions of topography, rock type, and structural alteration that otherwise proceed from west to east with predictable regularity. All of the coves have limestone or dolomite bottoms, and all are surrounded by mountains made up of much older and harder, less soluble, metamorphosed sandstones, siltstones, and conglomerates. How these older rocks, which would in a normal sequence occur far below the particular limestones and dolomites found in the cove bottoms, came to be higher is an important link in the geologic history of the mountains.

The basic geology of the Unaka Range has much to answer for, if all these oddities are to be explained. One simple fact stands out—the western ridges are formed from sedimentary rock, some layers of which are hard, others soft.

The beautiful coves of the Unakas, such as Cades Cove viewed here from Thunderhead Mountain, are geological oddities as well as scenic attractions. Cleared land in the bottom of the Cove rests on rocks that are younger than those of the surrounding mountains.

All the layers, hard and soft alike, have been squeezed and folded from their originally horizontal position to vertical or near vertical positions by forces pushing from the southeast, broken loose from their original positions, and pushed many miles toward the northwest. The rocks (which belong to what geologists call the Chilhowee Group) have been accordian-pleated, so to speak. Erosion, acting on these rock layers as the streams cut their way downward and extended their drainage basins headward, naturally removed the softer layers first. This left the hard-rock layers, the upward pleats of the analogy, standing up as ridges. The streams then became established on the soft-rock layers and cut deep gorges parallel to the direction of folding, establishing the trellis drainage so typical of the western part of the mountains.

In the middle part of the mountain ranges (proceeding toward the east) we again have sedimentary rocks, folded, compressed, and broken loose and shoved miles to the west. But the effect of erosion on these rocks has not been exactly the same as on rocks of the western part of the mountains. A major reason for the difference is the composition of the rocks. In general, in this part of the mountains, the rocks belong to a tremendously thick and very old sequence called the Ocoee Supergroup, which probably, even originally, were of more uniform hardness than were the rocks of the west. Additionally, however, the Ocoee rocks have been altered to a much greater extent by the changes in mineral content caused by heat and pressure applied over unimaginably long periods of time during the early history of the Unaka Range. This change in the nature of rock, called metamorphism, increases steadily toward the southeast, and has rendered even the softer layers much more resistant to erosion. As a result, these rocks of the middle ranges resist erosion much more uniformly than those to the west, and the drainage pattern consequently is much more uniform. Erosion acting on rocks of equal hardness would always produce a treelike drainage pattern.

The last stage in the west-to-east progression of rock types is represented by the granites and related rocks along the North Carolina line. Here the effects of heat and pressure have ''gone about as fur as they can go.'' Some of the rocks, in fact, appear to have originated as molten material (magma) from deep within the earth's crust. From here on east to the Blue Ridge and beyond in the Piedmont, the rocks are either igneous (once molten) or high-grade metamorphic rocks, some of which are so completely altered that it is almost impossible to tell whether they originated as sediments or not.

As mentioned before, all of the mountain mass has been broken loose from its original roots and shoved toward the northwest along a series of thrust faults that had an aggregate movement of many miles. This carried the mountain rocks far out over rocks like those of the Great Valley, and accounts for a number of otherwise puzzling features of the mountains today. The abrupt front of the range is at the point where the fault comes to the surface, where resistant rocks are pushed over the softer valley rocks. And the coves are no longer an anomalous interruption to the geologic progression when one realizes that the progression pertains only to the rocks above the fault, and the coves are places where erosion has cut a hole through the mountain mass and the fault below, to expose the valley rocks below the fault.

This also provides a way to know how far the mass has moved, or at least a minimum distance, since the rocks exposed in the cove bottoms are obviously part of the Great Valley sequence, demonstrating that those rocks extend at least that far beneath the thrust. Actually, there are other such "windows" through the overthrust, particularly the one near Grandfather Mountain in North Carolina, that suggest even greater movement than that based on the Tennessee coves.

The physical geography of the mountains and the basic rock types that make it up tell much about the history of the region and how it became what it is. The Unakas are a very ancient mountain range, with a history going back more than a billion years. There are gaps in the geologic record, periods of some tens of millions of years of which nothing is known, but all in all the evidence is fairly complete. The history, as recorded in the rocks, began over a billion years ago with the deposition of the oldest rocks, those that now make up the bedrock of the eastern part of the Unaka Range. Those earliest rocks were deposited as sediments on the bottom of a subsiding trough, a geosyncline like that in which the rocks of the Great Valley were deposited millions of years later, but far to the east of it. After many thousands of feet of sand, silt, and clay had been deposited, the geosyncline closed up, possibly squeezed shut by an early continental collision, and the sediments were compressed, folded, broken, and converted to hard rock. Some of the rocks were folded so deep into the crust of the earth that the natural heat generated by the decay of the minute quantities of radioactive minerals present in all rocks could not escape, and temperatures built up until parts of the mass melted and recrystallized. Other parts were invaded by liquid rock from still lower in the earth's crust, and the parts not melted or brought into contact with molten rock were nevertheless altered by the slow action of heat and pressure until they were changed into new forms, more durable under conditions of great pressure and high temperature. By this slow process of partial melting, recrystallization, and piecemeal alteration to new rock types, collectively called regional metamorphism, the sediments of the ancient geosyncline were converted to a mass of crystalline rock called the Carolina Gneiss. The Carolina Gneiss makes up the core of the Unaka Mountains, and is actually the same as the crystalline rock that underlies the sedimentary rock at great depth throughout Tennessee. In the Unakas, however, the crystalline rocks have been uplifted and faulted to a level so high that erosion has exposed them at the surface.

Following the metamorphism that produced the Carolina Gneiss, the region was uplifted, and erosion began to strip away the rocks and deposit them as sediments in a new geosyncline to the west. In the process a mountain range was formed that could be considered an ancestor to the present Unakas. The sediments filling the new geosyncline later hardened into the rock units called the Ocoee and Chilhowee groups (which now make up the main mass of the Unaka Range), about 600 million years ago, during the latest Precambrian and early Cambrian eras. A new period of folding and faulting followed, about 450 million years ago, during which some of the older rock units were thrust up and over younger rock units to the west. This was followed, about 375 million years ago, during the Devonian period, by another period of regional metamorphism, when heat and pressure not only altered the rocks of the newer geosyncline but also gave the Carolina Gneiss a new heat treatment. Still later these newly altered and realtered rocks were again subjected to folding and faulting, and more thrusts formed, moving masses of rock toward the northwest. It was while this later period of metamorphism and faulting was taking place that still a third geosyncline formed, again northwest of the old, this the one in which most of the rocks of the Great Valley were deposited. For the most part the rocks of this last geosyncline, although they have been converted to hard rock, have not been subjected to the heat and pressure that brings about metamorphism. There followed another great episode of folding and faulting, however, about 250 million years ago, during the Permian period, that created most of the structural complexities of the Great Valley and the Cumberland Plateau, and during which the whole area was lifted high above sea level, where it remains to this day.

During all the time since the last episode of faulting the mountains have been alternately eroded and uplifted, or perhaps both simultaneously, so that countless cubic miles of rock have been cut away in the process of carving the present mountains out of the roots of earlier ranges. At least three generations of peaks, maybe more, have stood where the Unakas stand today, each doubtless seeming a monument to the unchanging grandeur of creation, but each, in its turn, crumbling.

Whatever the exact sequence of events or processes that created the Unaka Range, most people will agree that the end product, with or without gold-bearing gravel, is a magnificent piece of scenery. So much so, in fact, that the Great Smoky Mountains are now part of the most visited of all the national parks. And with all the vacationers who come to admire and stay to spend money, it can still be said that "there's gold in them-thar hills!"

10. Explosion Structures: Tennessee's Mystery Craters

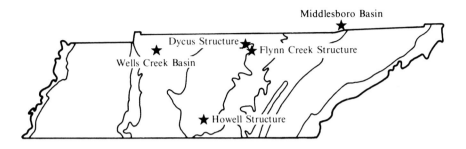

Middlesboro Basin

Dycus Structure

Flynn Creek Structure

Wells Creek Basin

Howell Structure

The pinpoint of light in the sky looked like a star at first, its growth slow and its movement almost imperceptible. But as growth and motion accelerated, the light brightened blindingly and filled the whole sky. No sound warned of its coming, for it approached far faster than the speed of sound. With shattering force the thing struck the ground, and an explosion ensued that shook half a continent. So great was the impact that the ground billowed and heaved, rock layers broke and recoiled away from the explosive center, and rocks thousands of feet below the ground were affected. When the debris of the explosion had settled, a crater miles across had formed, its raised rim encircling a deep depression, in the center of which stood an upraised hill. Beyond the rim great masses of material that had been ejected from the crater, from building-sized blocks to dust, radiated out from the center.

Such events have happened numberless times on the moon, and features like those described can be clearly seen through any good telescope. But this particular explosion took place not on the moon but on Planet Earth.

There are five structures in Tennessee, and many more in other states, that may have resulted from gigantic, near-surface explosions in the geologic past. The premier example in Tennessee is the Wells Creek Basin, on the south bank of the Cumberland River near Erin. Other structures, similar though smaller, are the Flynn Creek and Dycus structures in Jackson County, and the Howell structure in Lincoln County. Tennessee can claim only part of the fifth structure, the Middlesboro Basin, since only its outermost rim extends across the line from Bell County, Kentucky, into Claiborne County, Tennessee.

Each of these areas is roughly circular in outline, and all have geologic features in common that make geologists believe they have a common mode of origin. In every case there is a central area, commonly somewhat high topographically, in which all the rocks have literally been smashed, surrounded by a lower-lying area in which, as one goes away from the center, the rocks are less and less jumbled, though still badly disturbed. Surrounding this zone is an area of concentric rings of faulting, in which the rock layers have been tilted and raised or lowered, but are still recognizable as parts of particular rock formations. Radiating from the center of the structure are other faults, like spokes of a wheel, that offset not only the rock layers but also the concentric faults. Proceeding farther from center, the structural disturbance gradually dies away into the flat-lying, undisturbed rocks of the surrounding country. Perhaps the oddest fact of all about these structures is that each central hill is composed of pulverized fragments of rock formations that elsewhere in the area lie hundreds of feet below the surface.

Over the years a number of explanations have been advanced for the origin of these structures, all of which share one idea in common: that the formation of the structures involved explosive force. The controversy is basically over whether the explosion came from above or below, and what caused it.

The theory that the explosion came from below arose because early geologists recognized that rocks in the central part of the structures came from far below the surface. Something, apparently, had pushed them upward hundreds of feet. It was therefore proposed that molten volcanic rock from deep within the earth had come in contact with ground water, vaporized it to steam, and caused an explosion like the bursting of a gigantic boiler.

In more recent times, however, most geologists have come to believe that the structures were created by the impact of meteorites, either stony or metallic, or perhaps comets. This theory seems to be contradicted by the presence of the upraised central hill common to most of the structures, since it is difficult to see how an essentially downward-directed force could result in an upraised center. The explanation proposed is ingenious, and has the advantage of being based on a familiar phenomenon: the splash that occurs when a pebble is dropped into water or mud. The pebble momentarily displaces far more than its own volume of material, and when the liquid flows back into the void from all sides it meets in the middle and is deflected upward, resulting in the "splash." At first glance this explanation would not seem applicable to solid rock, but appearances, in this case, are deceptive. Geologists have long known that rocks, if confined and subjected to very high pressures, become plastic. If the meteorite (or comet) impact theory is correct, the explosive force of the impact instantaneously created such tremendous pressure in the rocks below that momentarily they became liquid, for all practical purposes.

The reader is perhaps wondering where in Tennessee there are craters like the ones described at the beginning of this chapter. On the moon, certainly; our astronauts have visited them, and on Mars, as the Mariner photos showed. But could one imagine anything less moonlike than the green, gently rolling countryside of Middle Tennessee? Tennessee's craters don't look like that anymore. To be sure, the structures are still round, or nearly so, and other evidence remains; but all were formed many millions of years ago, and the normal geologic forces of erosion and deposition have been hard at work. Like an adolescent diligently treating a case of acne, nature has smoothed the face of the land with the pumice of erosion, and when the scars proved too deep for that, plastered the make-up of younger sedimentary layers over them. Only now, as the present erosional cycle removes the superficial covering, can we see the outlines of the ancient structures revealed. The Wells Creek structure is probably the most completely exposed, as the Cumberland River has cut away whatever rock layers might once have covered it, leaving it thinly veiled with soil and river alluvium. The Flynn Creek, Howell, and

This drawing depicts a shatter cone from the central uplift in the Wells Creek structure. The apex of the cone points toward the source of the shock waves that formed it.

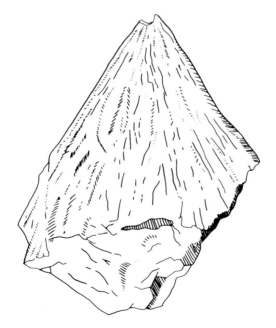

Dycus structures are only partly uncovered, mainly in those places where the valleys of small streams have cut down through the overlying rock layers and partly exposed them. The Middlesboro Basin was apparently once flooded by a fresh-water lake and is so thickly veneered with lake-bottom sediments that only a few exposures of shattered rock can be seen in the central part of the structure, plus remnants of the surrounding ring-faulting in the mountains surrounding the basin.

Evidence for the surface-explosion (meteorite/comet) explanation for these mysterious circular structures has been widely accepted, though still subject to being explained away by people (including many very capable geologists) who choose not to believe it. First, there are the shatter cones, strange conelike forms with fluted sides found in the intensely deformed cores of some of the structures. Laboratory experiments have shown that these cones can be formed only by subjecting rock to shock waves so powerful that there seems no rational way they could have been generated except by explosions. The points of the artificially created shatter cones point toward the source of the shock waves, but the rocks in which the natural cones are found are so jumbled up that it is hard to say which way they originally pointed. Careful mapping and plotting of shatter cone directions at Wells Creek, however, seem to indicate that they probably originally pointed upward to a "ground zero" at the surface as it existed at the time of the explosion.

Probably the most persuasive evidence for the above-ground explosion theory is that the structures have been reproduced in miniature by exploding shaped charges suspended above mud flats. The experimentally produced structures contain all the essential features of the crypto-explosive structures; circular shape, ring faults, depressed inner ring, and raised central hill, the whole schmear, so to speak.

An important fact about Tennessee's crypto-explosive structures is that they are all, apparently, of different ages. There is no way to date them exactly, but a relative idea of the age of each structure can be deduced because each must be younger than the youngest rock affected by its particular explosion and older than the oldest bed that stretches undisturbed across the top of it. In most instances this crude measuring system gives equally crude results. All we can say for certain about the age of the Wells Creek structure is that it is less than 330 million years old, since limestone beds of about that age are affected, and more than 70 million years old, since gravel beds of the Tuscaloosa Formation, deposited across the structure late in the Cretaceous period, are unaffected.

In other instances it is possible to be quite a bit more precise. The Flynn Creek structure is a very good example of this. At Flynn Creek the crater had not yet been smoothed away by erosion, which would surely have happened in a few thousand years, before the sea swept over the site and filled it with stinking black mud. Later that mud was compressed into black shale, part of a formation called the Chattanooga Shale that is so widespread it can be traced over most of the southeastern United States. Within the Flynn Creek structure the Chattanooga Shale is several times thicker than normal, showing that the crater was still deep when the sea rolled in. The Chattanooga Shale, because it contains a small amount of uranium, can be dated radiometrically and the Flynn Creek explosive event has thus been dated at about 360 million years ago, give or take a few million years.

Catastrophes are very popular literary subjects these days, and adding meteorite explosions to the list that already includes earthquakes, shipwrecks, burning skyscrapers, tidal waves, nuclear war, and so on, may seem to some the very last straw. What is the likelihood of another Wells Creek event in Tennessee? Fortunately, the odds against it seem pretty good. If the source of meteorites is cosmic debris left over from the formation of the universe, it follows that most of it would be swept up by the orbiting planets, or gravitate into the sun early in cosmic history. As the eons roll by there would be less and less of it, and the size would be smaller and smaller. Most of the fragments wandering through the solar system now are so small that when they cross the earth's path they burn up because of friction with the earth's atmosphere before reaching the ground. No meteorite big enough to duplicate any of the Tennessee craters has struck the earth in all of recorded history, a very encouraging fact, nor do astronomers warn of approaching disaster. So perhaps our battered planet has seen the last of these unwelcome visitors from outer space.

Selected Readings

Barr, T.C., Jr., 1961, *Caves of Tennessee*: Tenn. Div. Geology Bull. 64.

Faulkner, C.H., 1968, *The Old Stone Fort*: The University of Tennessee Press.

Floyd, R.J., 1957, *Rocks and Minerals of Tennessee*: Tenn. Div. Geology Inf. Circ. 5, 36 pp.

Fuller, M.L., 1912, *The New Madrid Earthquake*: U.S. Geol. Survey Bull. 494, 119 pp.

Hadley, J.B., and Goldsmith, R., 1963, *Geology of the Eastern Great Smoky Mountains, North Carolina and Tennessee*: U.S. Geol. Survey Prof. Paper 349 B, 118 pp.

Hardeman, W.D., Miller, R.A., and Swingle, G.D., 1966, *Geologic Map of Tennessee*: Tenn. Div. Geology, scale 1:250,000. Four sheets.

Killebrew, J.B., and Safford, J.M., 1874, *Introduction to the resources of Tennessee*: Tenn. Bureau of Agriculture First and Second Reports, 1193 pp.

King, P.B., 1964, *Geology of the Central Great Smoky Mountains, Tennessee*: U.S. Geol. Survey Prof. Paper 349 C, 148 pp.

————, and Ferguson, H.W., 1960, *Geology of northeasternmost Tennessee*: U.S. Geol. Surv. Prof. Paper 311, 136 pp.

————, Neuman, R.B., and Hadley, J.B., 1968, *Geology of the Great Smoky Mountains National Park, Tennessee and North Carolina*: U.S. Geol. Survey Prof. Paper 587, 23 pp.

Miller, R.A., 1974, *The Geologic History of Tennessee*: Tenn. Div. Geology Bull. 74, 63 pp.

Neuman, R.B., and Nelson, W.H., 1965, *Geology of the Western Great Smoky Mountains*: U.S. Geol. Survey Prof. Paper 349 D, 81 pp.

Rodgers, John, 1953, *Geologic Map of East Tennessee with explanatory text*: Tenn. Div. Geology Bull. 58, Part II, 168 pp.

Russell, E.E., and Parks, W.S., 1975, *Stratigraphy of the outcropping Upper Cretaceous, Paleocene, and Lower Eocene in Western Tennessee (includ-*

ing descriptions of younger Fluvial Deposits): Tenn. Div. Geology Bull. 75, 118 pp.

Stearns, R.G., 1958, *Cretaceous, Paleocene, and Lower Eocene geologic history of the Northern Mississippi Embayment area*: Tenn. Div. Geology Rept. Inv. 6, 24 pp.

_____, and Marcher, M.V., 1962, *Late Cretaceous and subsequent structural development of the northern Mississippi Embayment*: Tenn. Div. Geology Rept. Inv. 18, 8 pp.

Wilson, C.W., Jr., 1949, *Pre-Chattanooga stratigraphy in Central Tennessee*: Tenn. Div. Geology Bull. 56, 407 pp.

_____, 1958, *Guidebook to geology along Tennessee Highways*: Tenn. Div. Geology Rept. Inv. 5, 115 pp.

_____, 1962, *Stratigraphy and Geologic History of Middle Ordovician rocks of Central Tennessee*: Tenn. Div. Geology Rept. Inv. 15, 24 pp.

_____, Jewell, J.W., and Luther, E.T., 1956, *Pennsylvanian geology of the Cumberland Plateau*: Tenn. Div. Geology Folio, 21 pp.

_____, and Stearns, R.G., 1958, *Structure of the Cumberland Plateau, Tennessee*: Tenn. Div. Geology Rept. Inv. 8, 13 pp.

_____, 1968, *Geology of the Wells Creek structure, Tennessee*: Tenn. Div. Geology Bull. 68, 236 pp.

Index

(References to illustrations are in bold type)

Alluvium, 6, 10, 11, 26, 27
Anderson County, 58
Anticline, 57, **68**
Appalachian Mountains, 55, 76
Appalachian Plateau, 54–55
Arch. *See* anticline
Atlantic Coastal Plain, 67

Barrens, 44, 46
Bear Creek, 21
Bluegrass, 36, **38,** 46
Blue Ridge Mountains, 6, 20, 67, 75, 76
Bluff City, **49**
Bob's Bar well, 50, **51,** 52
Bradley County, **64**
Brown phosphate. *See* phosphate

Cades Cove, 78, **79,** 81
Cambrian period, 8–9, 11, 16, 52, 82. *See also*
  geologic time
Campbell County, 58
Cannon County, 39
Carolina Gneiss, 81, 82
Catoosa, 57
Caves, 44, 48, 50
Cenozoic era, 8–9. *See also* geologic time
Central Basin, 29, 32, 36–42, 44, 53
Charlie's Bunion, **77**
Chattanooga, 20, 55, 72
Chattanooga Shale, 47, 87
Chert, **33,** 44, 46–47
Chickasaw Bluffs, 5, 11, 14
Chilhowee Group, 80, 82
Chilhowee Mountain, 76
Chimney Top Mountain, **77**
Cincinnati Arch, 53
Claiborne County, 83

Cleveland, 63
Clinch Mountain, 72
Clingman's Dome, 76
Coal, 58
Coalings, The, 31
Coal mining. *See* coal
Coker Creek, 75
Comet. *See* meteorite
Conglomerate, 39
Continental drift, 67
Coon Creek, 12–13, 14
Coosa River, 21
Coves. *See* Cades Cove, Tuckaleechee Cove,
  Wear Cove
Crab Orchard Mountains, 57
Craters, 85
Cretaceous period, 8–10
Cross Mountain, 58
Crystalline rock, 81
Cumberland Block, 57, 60
Cumberland Caverns, 44, 48
Cumberland Gap, 54, **56**
Cumberland Mountains, 57, 58, **59, **60
Cumberland Plateau, 20, 35, 37, 44, 47,
  54–62, 63, 67, 72
Cumberland River, 29, 44, 55
Cyclora, 39

Dendritic drainage. *See* Drainage systems
Devonian period, 82. *See also* geologic time
Dolomite, 52, 78
Dome. *See* Nashville Dome, Ozark Dome
Douglas Reservoir, **68**
Drainage systems, 76, 78, 80
Duck River, 22, 23, 29, 43, 44, 55
Duck River Falls, **45**
Dycus explosion structure, 83, 85, 86

Earthquake, New Madrid. *See* New Madrid
  earthquake
Earthquakes in East Tennessee, 73–74
Eastern Highland Rim, 29, 34, 43–53
Elk River, 55
Elverton, 55
Embayment, Mississippi. *See* Mississippi Em-
  bayment
Emory River, 57
Epicenter, earthquake, 6
Era. *See* geologic time
Erin, 83
Escarpments, 46, 48, 55, 60, 62
Explosion stuctures, 83–87

Fall Creek Falls, 62
Falls. *See* waterfalls
Faults, 55–57, 60, 63, **64,** 73, 75, 80, 81; as
  cause of New Madrid earthquake, 10; four
  principal kinds of, **66**
Flood Plain of the Mississippi River, 3–11, 14,
  16, 18, 23, 26
Flynn Creek, 83, 85, 86, 87
Fort Payne Formation, 20, 44, **45,** 46–47
Fossils, 26, 27, 36, 37, 39, **42**
Fossil-collecting, 12–13
French Broad River, 76
Furnaces, 28, **33**

Gatlinburg, 63, 76
Geologic regions of Tennessee, vii, viii–ix
Geologic time, 6–10
Geosyncline, 69, 70, 73, 82
Glaciers, 10–11, 18
Glauconite, 34
Gold mining, 75
Gorges, Pleistocene, 11
Grand Canyon of the Tennessee River, 20
Grandfather Mountain, 81
Granite, 41, 65, 80
Great Smoky Mountains, 63, 65, 66, 75, 76
Great Valley of the Appalachian region, 67
Great Valley of East Tennessee, 55, 63–74, 76
Great Western furnaces, **33**
Gulf (of Mexico) Coastal Plain, 26, 41, 70
Gunpowder, 44
Guntersville, Ala., 20

Hardeman County, 24
Hemotite, 29
Hickman County, **30**
Higginbotham Cave. *See* Cumberland Caverns
Highland Rim, 37, 39, 47, 55

Hiwassee River, 76
Holston Formation, 70
Howell explosion structure, 83, 85, 86

Ice age. *See* Pleistocene
Igneous rock, 41, 80
Iron Mountains, 75
Iron ore, 29, 31, 34–35; industry, 28, **30;** min-
  ing, 28

Jackson County, 83
Jackson well, 50, 52
Jefferson County, **68**
Joint, 65
Jurassic period, 73. *See also* geologic time

Karst topography, 48
Kentucky, 16, 36, 60
Kentucky Lake, 21, **22,** 23
Knees in cypress trees, **4,** 5
Knox County, 63
Knoxville, 60, 70, **71**

Lake City, 60
Lakes, man-made. *See* man-made lakes
Lakes, oxbow. *See* oxbow lakes
Levees, natural, 5
Lexington, Ky., 36
Lignite, 58
Limestone, 36, 37, 39, 47, 48, 52, 60, 70, 78
Limonite, 29, 31, 32, **33,** 35
Lincoln County, 83
Little Duck River, 43, 44
Little Emory River, 55–57
Little Tennessee River, 76
Loess, 11, 14, **15,** 16–18
Lone Mountain, 60
Lookout Mountain, 60, 63
Louisiana, 16

McMinnville, 48
McNairy County, fossil collecting in, 12–13
Madison County, 16–17
Magma, 80
Man-made lakes, 21
Marble, 70, **71**
Mariner (interplanetary missile), 85
Mars, 85
''Master fault'' (*décollement*), 65
Meander scars, 5, **7**
Memphis, 5, 14, **15**
Mesozoic era, 8–10. *See also* geologic time
Metamorphic rock. *See* granite

Metamorphism, 70, 80, 81, 82
Meteorite, 84
Mexico, Gulf of, 6, 10, 18, 20, 21, 78
Middlesboro Basin, 83, 86
Mississippian period, 9, 16. *See also* geologic time
Mississippi Embayment, 10, 16, 21, 26.
Mississippi River, 5, 10, 11, 16, 70, 78. *See also* Flood Plain of the Mississippi River
Missouri, 16
Mobile Bay, 21
Molten rock. *See* magma
Monroe County, 75
Monteagle, 54
Montgomery Bell, 28
Moon, 85
Morgan County, 57–58
Morrill Cave, **49**
Mosasaur, 13, **13**
Mount Mitchell, 76
Murfreesboro, 36
Muscle Shoals, 20, 21

Nashville, 36, 46, 69
Nashville Dome, 8, 10, 16, 29, 34, 36, 37, 39, 41, 53
New Madrid earthquake, 3, 5, 10
New York (state), 54
Niter, 44, 50
Normal fault. *See* faults, four principal kinds
North Carolina, 6
Northern Highland Rim, 29
Nunnelly, **30**

Ocoee River, 76, 82
Ocoee Supergroup, 80
Ohio River, 10, 23, 78
Oil, 50–52; fields, 44; rig, **51**
Okefenokee Swamp, 58
Old Stone Fort, 43, **45, 47**
Ooltewah, **64**
Ordovician rock, 41
Oxbow lakes, 5, **7**
Ozark Dome, 8, 10, 16, 53

Paducah, Ky., 20
Paleozoic era, 8–9, 24, 53, 55, 72. *See also* geologic time
Parksville, **68**
Pascola Arch, 8, 10, 16
Peat, 58
Pennyroyal Plateau, 29
Permian period, 82. *See also* geologic time
Period. *See* geologic time

Phosphate, 37, 39
Pickett State Park, **61**
Piedmont, 67
Pigeon River, 76
Pine Mountain, 60, block, **56**
Plateau Slope of West Tennessee, 12–18, 23, 26
Pleistocene, 10, 18, 23, 27. *See also* geologic time
Plesiosaur, 12, 13
Polar ice cap, 18
Polk County, **68**
Powell Mountain, 60
Precambrian era, 8–9, 82. *See also* geologic time
Putnam County, 24

Raccoon Mountain, 60
Reelfoot Lake, 5–10
Regional metamorphism. *See* metamorphism
Reverse fault. *See* faults, four principal kinds
Ridges of East Tennessee, 72
Roane County, 55, 56
Rock, 6, 14, 72, 78, 80, 139; folding of, 55
Rocky Mountains, 6

St. Louis Limestone, 47–48
Sand hills, 14
Sand Mountain, 60
Sandstone, 39, 60, **61,** 62
Scott County, 58
Sea-floor spreading, 67, 70
Sedimentary rock, 69, 80, 81
Sequatchie Valley, 20, 57, **59**
Sevier County, 63
Shale, 60, 62
Shatter cone, **85,** 86
Short Mountain, 38–39, **40,** 53
Sinkholes, 48, 50
Smelting, 28
Smoky Mountains. *See* Great Smoky Mountains
Southern Highland Rim, 29
Spencer, 57
Spring Creek oil field, 50, 52
Stalactites, 50
Stalagmites, 50
Stone Mountains, 75
Stewart County, **33**
Suck, The, 19, 20
Syncline, **68**

Tear fault. *See* faults, four principal kinds
Tellico River, 76

Tennessee River, 6, 8, 14, 27, 55, 72, 78;
  peculiarities of flow, 19–23; valley of, 23
Tennessee-Tombigbee Waterway, 27
Tennessee Valley Authority (TVA), 21
Tertiary period, 26. *See also* geologic time
Thrust fault. *See* faults, four principal kinds
Thunderhead Mountain, **79**
Tombigbee River, 21
Trellis drainage. *See* Drainage systems
Tuckaleechee Cove, 78
Tuscaloosa Formation, 86

Unaka Mountains, 19, 67, 72, 75–82
Unconformity, 23, 24, **25,** 26

Unicoi Mountains, 75

Van Buren County, 62
Vicksburg, Miss., 18
Virginia, 20, 60

Walden, Eligah, 57
Walden Ridge, 19–20, 55–57
Waterfalls, 44, 47, 62
Wayne County, 35
Wear Cove, 78
Wells Creek Basin, 83, 85–86
Western Highland Rim, 26, 28–35, 52
Western Valley of the Tennessee River, 19–27
"Windows" (coves), 81